PORTRAITS OF
TREES OF HONG KONG
AND SOUTHERN CHINA

Illustrations by
Sally Grace Bunker

Text by
Richard M. K. Saunders
and Chun-Chiu Pang

EARNSHAW
BOOKS

Portraits of Trees of Hong Kong and Southern China

Illustrations by Sally Grace Bunker

Text by Richard M. K. Saunders and Chun-Chiu Pang

ISBN-13: 978-988-8552-03-0

SCIENCE / Life Sciences / Botany

EB109

Published by Earnshaw Books Ltd. (Hong Kong)

Contents

—— ☙ ——

Foreword (by Professor Stephen Blackmore)   7

Preface   9

Acknowledgements   12

Introduction

The tree flora of Hong Kong   13
The evolutionary diversity of trees   16
Botanical nomenclature   19
The art of botanical illustration   20

Plant portraits

Acacia auriculiformis Cunn. ex Benth.   24
Acacia confusa Merr.   26
Acer sino-oblongum F. P. Metcalf   28
Acronychia pedunculata (L.) Miq.   30
Adenanthera microsperma Teijsm. & Binn.   32
Adina pilulifera (Lam.) Franch. ex Drake   34
Adinandra millettii (Hook. & Arn.) Benth. & Hook. f. ex Hance   36
Alangium chinense (Lour.) Harms   38
Aleurites moluccana (L.) Willd.   40
Anneslea fragrans Wall. var. *hainanensis* Kobuski   42
Annona squamosa L.   44
Aporosa dioica (Roxb.) Müll. Arg.   46
Aquilaria sinensis (Lour.) Spreng.   48
Archidendron lucidum (Benth.) I. C. Nielsen   50
Ardisia quinquegona Blume   52
Artocarpus hypargyreus Hance ex Benth.   54
Aucuba chinensis Benth.   56
Bauhinia purpurea × *variegata* 'Blakeana'   58
Bombax ceiba L.   60
Bruguiera gymnorhiza (L.) Savigny   64
Callicarpa nudiflora Hook. & Arn.   66
Camellia crapnelliana Tutch.   68
Camellia hongkongensis Seem.   70
Castanopsis fissa (Champ. ex Benth.) Rehder & E. H. Wilson   72
Casuarina equisetifolia L.   74
Celtis sinensis Pers.   76
Cerbera manghas L.   78

Cinnamomum camphora (L.) J. Presl   80

Cratoxylum cochinchinense (Lour.) Blume   82

Cyclobalanopsis championii (Benth.) Oerst.   84

Dalbergia assamica Benth.   86

Daphniphyllum calycinum Benth.   88

Delonix regia (Bojer) Raf.   90

Dimocarpus longan Lour.   92

Diospyros morrisiana Hance   94

Elaeocarpus chinensis (Gardner & Champ.) Hook. f. & Benth.   96

Endospermum chinense Benth.   98

Engelhardia roxburghiana Lindl. ex Wall.   100

Enkianthus quinqueflorus Lour.   102

Eriobotrya japonica (Thunb.) Lindl.   104

Exbucklandia tonkinensis (Lec.) H. T. Chang   106

Ficus variolosa Lindl. ex Benth.   108

Garcinia oblongifolia Champ. ex Benth.   110

Glochidion zeylanicum (Gaertn.) A. Juss.   112

Gmelina chinensis Benth.   114

Heritiera littoralis Aiton   116

Hibiscus tiliaceus L.   118

Homalium cochinchinense (Lour.) Druce   120

Ilex rotunda Thunb.   122

Illicium angustisepalum A. C. Sm.   124

Itea chinensis Hook. & Arn.   126

Kandelia obovata Sheue, H. Y. Liu & J. W. H. Yong   128

Leucaena leucocephala (Lam.) de Wit   130

Liquidambar formosana Hance   132

Lithocarpus glaber (Thunb.) Nakai   134

Litsea cubeba (Lour.) Pers.   136

Litsea glutinosa (Lour.) C. B. Rob.   138

Livistona chinensis (Jacq.) R. Br. ex Mart.   140

Lophostemon confertus (R. Br.) Peter G. Wilson & J. T. Waterh.   144

Macaranga tanarius (L.) Müll. Arg.   146

Machilus chekiangensis S. K. Lee   148

Machilus velutina Champ. ex Benth.   150

Magnolia championii Benth.   152

Mallotus paniculatus (Lam.) Müll. Arg.   154

Melaleuca cajuputi Roxb. subsp. *cumingiana* (Turcz.) Barlow   156

Microcos nervosa (Lour.) S. Y. Hu   160

Myrica rubra (Lour.) Sieb. & Zucc.   162

Ormosia emarginata (Hook. & Arn.) Benth.   164

Osmanthus matsumuranus Hayata   166

Paliurus ramosissimus (Lour.) Poir.   168

Pandanus tectorius Parkinson   170

Pavetta hongkongensis Bremek.   174

Pentaphylax euryoides Gardn. & Champ. 176

Photinia benthamiana Hance 178

Phyllanthus emblica L. 180

Pinus elliottii Engelm. 182

Pinus massoniana Lamb. 184

Podocarpus macrophyllus (Thunb.) Sweet 186

Polyspora axillaris (Roxb. ex Ker Gawl.) Sweet 188

Pyrenaria spectabilis (Champ.) C. Y. Wu & S. X. Yang 190

Pyrus calleryana Decne. 192

Reevesia thyrsoidea Lindl. 194

Rhaphiolepis indica (L.) Lindl. ex Ker 196

Rhodoleia championii Hook. 198

Rhus chinensis Mill. 200

Sapium discolor (Champ. ex Benth.) Müll. Arg. 202

Saurauia tristyla DC. 204

Schefflera heptaphylla (L.) Frodin 206

Schima superba Gardner & Champ. 208

Scolopia chinensis (Lour.) Clos 210

Sinosideroxylon wightianum (Hook. & Arn.) Aubrév. 212

Sloanea sinensis L. 214

Sterculia lanceolata Cav. 216

Styrax suberifolius Hook. & Arn. 218

Symplocos congesta Benth. 220

Syzygium jambos (L.) Alston 222

Tetradium glabrifolium (Champ. ex Benth.) T. G. Hartley 224

Thespesia populnea (L.) Sol. ex Corr. 228

Toxicodendron succedaneum (L.) O. Kuntze 230

Trema tomentosa (Roxb.) Hara 232

Turpinia montana (Blume) Kurz 234

Vernicia montana Lour. 236

Viburnum odoratissimum Ker Gawl. 238

Zanthoxylum avicennae (Lam.) DC. 240

Glossary 243

References 250

Index of Plant Names 271

Index of Chinese Plant Names 280

Subject Index 282

About the Authors 289

Foreword
by Professor Stephen Blackmore

———— ❧ ————

The tree which moves some to tears of joy is in the eyes of others only a green thing which stands in the way. Some see Nature all ridicule and deformity, and by these I shall not regulate my proportions; and some scarce see Nature at all. But to the eyes of the man of imagination, Nature is Imagination itself. As a man is, so he sees.

William Blake, *Letters* (1906)

This inspiring book will open the eyes of everyone who reads it both to the beauty of trees and to their profound importance. Growing quietly in the background of our lives, trees shape the terrestrial environment, providing habitats for more than half the animal species on land. They replenish the oxygen in the air we breathe, lock away carbon and support the year-round flow of streams and rivers. Beyond this they provide a myriad useful products, from fruits and forage to medicines, resins and timber.

It may come as a surprise that Hong Kong is rich in trees given that it is one of the most densely populated places on earth, being home to some 7.4 million people, about 0.1% of humanity. Remarkably, Hong Kong has 390 indigenous species of trees and many more that have been introduced for one purpose or another. New species and varieties continue to be discovered. As recently as 2014 a species of Hornbeam, *Carpinus insularis*, native to Hong Kong and new to science was described from a tiny population found by botanists at Violet Hill. It has already been propagated and bought into cultivation so that the wild populations can be reinforced and protected for the future. Further discoveries may well follow, especially as the eyes of more people are opened to the diversity and value of trees and forests.

Hong Kong is, in fact, so rich in trees that recognising and distinguishing between them is a significant challenge. This book will help overcome this obstacle and enable more people to discover the joy of knowing trees. When, as a budding botanist, I lived in Hong Kong in the 1960s there were few sources to turn to for help. *The Hong Kong Countryside* by Geoffrey Herklots was a treasured possession and a source of insight into some of the trees and much else besides. Fortunately, today the situation is very different and naming plants is much easier thanks to numerous highly accessible photographic guides and the highly professional *Flora of Hong Kong* published in four volumes by the Agriculture, Fisheries & Conservation Department of the Hong Kong Government.

Important though it is, as a starting point, to be able to identify and name a tree, readers with an appetite for knowledge will want more and this book will satisfy their hunger. This is due in large part to the distinctive and complementary expertise of the three authors who have worked so creatively to produce a new body of work with great strengths. Firstly, Sally Bunker's watercolour paintings provide an exquisite example of botanical art. This is a discipline which requires not only a good eye and a steady hand but the patience to observe and document each specimen many times, through the seasons.

Some life stages will sit patiently, allowing ample time for their portraits to be painted; others are transient and must be captured in a fleeting moment. A good botanical artist can capture the essence of a plant, drawing together its key aspects and features in ways photographs cannot truly replicate. Secondly, this book tells the stories of just over a hundred carefully selected trees chosen to cover the variety of habitats present in Hong Kong and the surrounding region while introducing the diversity of plant families and their strategies for survival. While the scope is quite different from Neil MacGregor's *A History of the World in 100 Objects*, there are parallels. The texts for the selected trees encompass the history of botanical and geographical exploration, the discovery and distribution of useful trees around the world, their classification, ecology and numerous cultural associations from the culinary to the medicinal. The stories illuminate the wonderful ways trees overcome the challenge of being rooted to one spot: supressing the growth of competing neighbours, exchanging genes with sometimes distant members of their species and dispersing their seeds and fruits to new and suitable locations. Trees are revealed as active participants not passive components in the world around us and rightly celebrated for their beauty and individuality.

It may be inappropriate to pick a favourite from such a feast, but the magnificent flowers of the cotton tree (*Bombax ceiba*) take me back to 1963 and the flat on Shiu Fai Terrace where cotton trees brushed against our bedroom windows. Their branches buzzed with the relentless and indescribable sound of cicadas and later came the magic of the cotton bursting from the fruits to cover the ground. Those trees were felled, soon after, to make way for new development and, at the time, it seemed perfectly possible that the expansion of the city and growth of skyscrapers might one day eradicate trees from Hong Kong. For, as the Introduction describes so vividly, accounts by western naturalists in the eighteenth and nineteenth centuries painted a bleak picture, echoed in Lord Palmerston's derisive description: 'A barren rock with nary a house on it. It will never be a mart for trade.' Happily, the importance of forests for what we now call 'ecological services' led to the establishment of the Country Parks and today, thanks to this and other changes in land management, Hong Kong has extensive forests. To me, this remarkable transformation provides a hopeful and much-needed example to the world. We now know how, with sufficient effort and expertise, to restore the ecological health of degraded landscapes. Human enterprise and endeavour need not be synonymous with the destruction of nature, even in one of the world's great centres of trade. Too often in the past we failed to understand that we ourselves are an integral part of nature, but now we are alive to the reality that we can no more thrive in a world without nature than our heads can live without our bodies.

Those already moved to tears of joy by trees will delight in this book and even those who see trees as green things that stand in the way are unlikely to be unmoved by it. I wonder what would Lord Palmerston say if presented with a copy?

Professor Stephen Blackmore CBE VMH FRSE
Queen's Botanist and Honorary Fellow, Royal Botanic Garden Edinburgh
Chair of Botanic Gardens Conservation International

Preface

Flowers do not fill a life, but they furnish large corners, which might otherwise have been occupied by cobwebs; and while studying them the thinker is brought into contact with his own kind; and sometimes they tell him stories…

John Trevena, *Adventures Among Wild Flowers* (1914)

Trees are a conspicuous component of Hong Kong's cityscape and country parks: in addition to the 390 tree species that are indigenous to the territory[1] there are numerous exotic species that have been introduced in plantations or as ornamentals in public gardens and along roadsides. Despite their physical dominance, both in terms of size and number of individuals, trees can easily be overlooked and ignored.

The objective of this book is to draw attention to our local tree flora, using illustrations and accompanying text to tease out individual narratives: each species of tree has an associated 'story' with contrasting focus. In many cases these narratives are human-focused, ranging from the history of their discovery, naming and cultivation (*e.g.*, *Bauhinia purpurea* × *variegata* 'Blakeana'), to their uses, often as plantation trees (*e.g.*, *Pinus elliottii*), fire-breaks (*e.g.*, *Acacia confusa*), ornamentals (*e.g.*, *Bombax ceiba*), and as a source of food (*e.g.*, *Dimocarpus longan*) or medicines (*e.g.*, *Alangium chinense*). Human activities have also had a profound impact on our tree flora as a result of unsustainable logging and habitat destruction, and conservation issues are consequently a recurrent theme in this book. The accidental introduction of the pinewood nematode, *Bursaphelenchus xylophilus*, to Hong Kong in the late 1970s is a cogent example of the ecological damage incurred by alien species: the nematode-borne 'pine wilt disease' has devastated populations of our only native pine species, *Pinus massoniana*, and has had a profound and long-term impact on the local landscape.[2,3]

As might be expected, many of the narratives in this book focus on plant structure, particularly flower and fruit morphology. In his seminal book *On the Origin of Species*[4] Charles Darwin wrote of the 'struggle for existence', in which organisms compete against each other and their environment to survive and reproduce. This struggle for existence is as relevant to plants as it is to animals and has been the driver of evolutionary change through natural selection. It is not possible to understand the enormous diversity of floral structure without appreciating the ways in which the flowers are pollinated and—in the case of animal-pollinated species—the features of the flowers that attract and reward floral visitors. Wind-pollinated species such as *Liquidambar formosana* inevitably have inconspicuous flowers, in marked contrast with the often flamboyant floral displays of animal-pollinated species; the latter furthermore exhibit an enormous diversity of structures associated with different pollinators, including bats, birds and insects such as beetles, thrips, flies, bees and butterflies. In some cases the co-evolutionary specialisation of flowers and their pollinators achieves such an extreme degree of specificity that a mutual inter-dependence develops, in which the two species cannot survive or reproduce without the involvement of the other. Fig trees (*e.g.*, *Ficus variolosa*) are perhaps the best example of such a complex 'mutualistic' relationship: each *Ficus* species is typically pollinated by a single species of fig wasp (belonging to the family Agaonidae) and co-evolution has resulted in extreme morphological specialisation in both the fig and wasp.

Another focus of the narratives in this book are the various reproductive strategies that have evolved in plants to increase genetic diversity by promoting breeding between different individuals, rather than within the same flower or between different flowers on the same individual. In some species this has been achieved by separation of sexes (either 'dioecy', as in *Macaranga tanarius* and *Pandanus tectorius*, in which individual trees are either male or female; or 'monoecy', as in *Castanopsis fissa* and *Liquidambar formosana*, in which separate male and female flowers are borne on the same tree). Alternative evolutionary strategies to promote breeding among different individuals include temporal changes, with staggered maturation either of flowers of different sexes within an individual (*e.g.*, *Dimocarpus longan*) or of the male and female organs within the same flower (*e.g.*, *Delonix regia* and *Schefflera heptaphylla*), and differential movement, wilting or drop of floral organs over time (*e.g.*, *Delonix regia*).

The morphological diversity of flowers has a consequential effect on the structure of the fruits that develop from the flower after fertilisation: although individual flowers typically develop directly into single fruits, in some species (*e.g.*, *Liquidambar formosana* and *Pandanus tectorius*) clusters of flowers develop into a single multiple fruit, whereas in others (*e.g.*, *Acer sino-oblongum*) individual flowers develop into several distinct fruit units at maturity. Diverse fruit and seed morphologies have also evolved in response to different dispersal mechanisms, with adaptations that promote transport of propagules by wind (*e.g.*, *Polyspora axillaris*) and water (*e.g.*, *Hibiscus tiliaceus*), and, in the case of animal-dispersed fruits, adaptations (most commonly sugar-rich, fleshy fruits) that attract different fruit-eating animals (frugivores).

Research into plant diversity and evolution is an active and vibrant scientific endeavour, typically incorporating a broad range of data sources. Contemporary research often includes a comparative study of molecular variation in the organism's DNA, which is the carrier of genetic information in cells. Random mutations in the DNA (in the form of nucleotide substitutions) accumulate over time so that species that diverged earlier in their evolutionary history are likely to have accumulated a greater number of differences. DNA sequencing technology now enables these differences to be compared and used to reconstruct the evolutionary tree (phylogeny) of the taxonomic group, showing the order in which divergence events occurred. If the phylogeny is appropriately calibrated using fossils of known age, it is furthermore possible to estimate the timing (in millions of years) since the evolutionary divergence is likely to have occurred. This type of research has resulted in some major changes in our understanding of the classification of plants: in the context of Hong Kong's tree flora, these changes include the recognition that the common local 'Gordonia' species does not belong to the genus *Gordonia* itself (which is now recognised as exclusively North American), but should instead be recognised in the genus *Polyspora*, as *Polyspora axillaris*.[5] DNA sequence data is also widely used to explain the evolutionary biology and geographical distributions of organisms: *Liquidambar formosana*, for example, belongs to a genus that exhibits an intriguing East Asian-North American biogeographical disjunction that resulted from dispersal from Asia into North America via land bridges during the Late Middle Miocene thermal maximum (17–15 million years ago).[6]

The discussion of these different narratives is presented in an ordered way in this book, adopting the scientific nomenclature that has been used as an international standard for over two and a half centuries. The plant species names used here are largely consistent with those adopted in the four-volume *Flora of Hong Kong*,[7] published by the Agriculture, Fisheries & Conservation Department of the Hong Kong Government. In many cases, species have received multiple names (either in error due to duplicate descriptions, or because of differing taxonomic opinion). We have not attempted to present a comprehensive list of such synonyms, but instead have restricted ourselves to names that have commonly been used in the regional literature, including: the early *Flora Hongkongensis*,[8] published by George Bentham in 1861; Henry F. Hance's supplement[9] to Bentham's book, published in 1872; and the *Flora of Kwangtung and Hong Kong (China)*,[10] published by Stephen T. Dunn and William J. Tutcher in 1912. In some cases we deviate from the plant family names used in the *Flora of Hong Kong*, preferring instead to adopt the *Angiosperm Phylogeny Group* classification[11-14] that is consistent with current understanding of evolutionary diversification. The Latin names and citation methods may be confusing but are nevertheless essential for effective communication; the Introduction to this book includes an overview of the internationally accepted rules and regulations governing the application of plant names.

The watercolour paintings in this book have been based on direct observation of living trees in Hong Kong, and as far as possible in their native habitat. This has necessitated repeat visits to each tree to ensure that structures that develop in different seasons, such as flowers and fruits, can all be drawn from the same individual. In many cases, the illustrations include individual flowers and fruits, dissected to reveal their constituent organs, including the sepals, petals, stamens and carpels from the flowers and the seeds within the fruit. We have not made any attempt at uniformity of organs represented in each plate since the features of interest vary greatly between species. The illustrations include a black-and-white pencil sketch showing the overall architecture of the tree, revealing the often-characteristic branching pattern and crown shape. The bark morphology is also represented by the inclusion of circular drawings based on direct pencil 'rubbings' of the trunk.

We hope that the illustrations and accompanying text will stimulate interest in our local trees. We have endeavoured to ensure that all facts and hypotheses included in this book are fully cross-referenced with the relevant scientific literature so that readers who wish to pursue their interest further may do so. Although we have tried to avoid relying too heavily on specialised botanical terminology, its use is sometimes unavoidable when trying to convey information unambiguously; we have accordingly included a glossary of essential scientific terms.

Richard Saunders
Chun-Chiu Pang

Acknowledgements

———— ∾ ————

We are deeply indebted to The Swire Group Charitable Trust for helping subsidise the production costs of this book, and to Tina Chan in particular for her support.

We are grateful to Bill Flanz and Markus Shaw of WWF Hong Kong, who helped kick-start the project by introducing Sally to the other authors, and Sally's husband, Bob Bunker, for his unwavering support and encouragement throughout the project. We are also very grateful to Prof. Ron Hill and two anonymous reviewers for their comments, which have greatly improved the text, and Prof. Steve Blackmore for his generosity in contributing such an eloquent Foreword. Also Paul Melsom, for his help in finding trees, and Brian Tilbrook for all his encouragement to Sally in her art work.

Sally would specifically like to thank Lorette Roberts for guiding her into a new career as a botanical artist, the UK Society of Botanical Artists, and Clive Rigby, and William and Edward Bunker for their support.

Richard and Chiu are indebted to their wives, Sue and Vivienne, for their continuous support.

Introduction

—— ❦ ——

The Tree Flora of Hong Kong

> In the morning we weighed anchor and steered to the Chinese coast … We had Lantoa [Lantau] on our right and the Southern isles of Limes [the Lema Islands] on the left: the sea formed high billows rolling in from the isles, which were quite green with plants, but had no woods.
>
> Pehr Osbeck, *A Voyage to China and the East Indies* (1757)

On 22nd August 1751, Pehr (Peter) Osbeck, naturalist and chaplain aboard the Swedish ship *Prins Carl*, sailed into Hong Kong waters. His description of the flora (quoted above from his book *A Voyage to China and the East Indies*, published in Swedish in 1757 and translated into English 14 years later[15]) is the earliest known account of Hong Kong's vegetation by a western naturalist. Osbeck makes a startling observation: although the islands were clearly verdant *they were devoid of trees*. From the wording used by Osbeck, it seems likely that he was describing the vegetation on the Lema Islands—which lie 11 miles due south of Hong Kong Island and are not to be confused with Lamma—rather than Lantau. The French landscape and botanical painter Jean Louis Prévost had previously visited the Lema Islands in 1742 and had similarly described them as 'sterile and covered with rocks,' although his description and its accompanying panoramic engraving were not published until 35 years after the visit.[16] Considerable caution is necessary before extrapolating these observations to other islands in the region, although naturalists who visited Hong Kong in the nineteenth century appear to have been similarly unimpressed.

Richard Brinsley Hinds, naturalist aboard *H. M. S. Sulphur* (1836–42), visited Hong Kong Island in January and February 1841 and described it as 'wild, dreary, bleak, and apparently barren.'[17] He reported 'large bare masses of rocks unscreened by foliage' but conceded that the flora was more diverse in the west of the island, where 'some stunted pines try to assume the importance of trees.' He regarded the distant view of Hong Kong as presenting 'a picture of sameness and barrenness not likely to convey a very favourable impression of the variety and interest of the vegetation,' but noted that the greatest diversity was to be found in valleys, where, although 'trees can scarcely be said to exist … there is a great variety, in sheltered situations, of low pretty evergreen shrubs …'[17]

A similarly bleak picture of Hong Kong's vegetation was portrayed by Berthold Seemann, a German naturalist who circumnavigated the globe on *H. M. S. Herald* (1845–51) and visited the territory in 1850: 'To a stranger landing, or regarding the island from the sea, the aspect of Hongkong is very unpromising, conveying the idea of almost absolute sterility. The hills are covered by a mantle of coarse grass, amidst which rise masses of bare, blackened rocks; while the monotonous scene seems only varied by a few bushes, or a solitary tree, studded here and there, and by scattered groves of the *Pinus Sinensis* [*Pinus massoniana*] clothing some of the declivities.'[18] Seemann suggested that the impoverished tree flora of Hong Kong was due to human activities and that the Chinese red pine, *Pinus massoniana*, 'was at one time far more common, and originally formed dense woods on the flanks of the hills of all the islands' but had decreased due to extensive burning by the local population.[18]

Although there is archaeological evidence for human settlement in Hong Kong dating from 4000–2500 BC,[2] it is unlikely that these early populations were large enough to have had a significant impact on the tree flora. There is evidence of lime kilns dating from 300–900 AD,[2] however, in which locally gathered shells and coral were fired to produce mortar for export further inland; these kilns would undoubtedly have been fuelled using local timber. Increased occupation of Kowloon and the New Territories occurred from the tenth century, with more extensive records from the fourteenth century; Cantonese clan lineages from northern Guangdong subsequently established large populations from the seventeenth century, as evidenced by recorded genealogies as well as archaeological relics.[2] In addition to the Cantonese, who typically cultivated the fertile and productive valleys and lowland areas, Hong Kong was settled by other ethnic groups, including later Hakka immigrants, who cultivated upland areas. Cultivation clearly extended to the highest hills, even predating the arrival of the Hakka: there is evidence that the higher slopes of many hills, including Tai Mo Shan, Castle Peak and Lantau Peak, had already been terraced for tea cultivation in the seventeenth century[2] (and in the case of Tai Mo Shan, possibly even as early as the thirteenth century[19]).

It is difficult to assess the degree of human impact on the tree flora of Hong Kong prior to the cession of the territory to the British in 1841, largely due to uncertainties regarding population size. Correspondence by Lieutenant Thomas B. Collinson of the Royal Engineers, dating from 1843–45, suggests that there were approximately 11 villages on Hong Kong Island, with *c*. 400 acres of associated land under cultivation;[20] the largest of these villages, Aberdeen and Stanley, are unlikely to have exceeded more than 1,000 inhabitants each.[20] It is also difficult to assess anthropogenic impact on the forests of Hong Kong at this time because of problems in estimating the wood or charcoal requirements of the population. Although wood is necessary for boat and house construction and for heating during the winter, fires for food preparation often have to generate a brief, intense heat and in these cases would presumably have been fuelled using kindling rather than larger logs. Despite these uncertainties, it is likely that the increasing population and associated agriculture and deforestation from the seventeenth century onwards contributed to the degraded forests observed by Osbeck, Hinds, Seemann and other naturalist-explorers in the eighteenth and nineteenth centuries.

It has been estimated that only 3–4% of Hong Kong Island was forested by the 1840s.[2,21] Two substantial forests persisted in Happy Valley and 'Little Hong Kong' (near Aberdeen), with a smaller area at Tai Tam Tuk.[21] Early botanical collections reveal that Happy Valley was floristically very rich, with many Fagaceae species, and may have represented a remnant of the original forest cover; this forest was cleared around 1900, however.[21] The flora of Little Hong Kong was less diverse, and although largely cleared in 1922 for a camphor wood plantation, a small remnant survives north of Nam Fung Road and has been recognised as a Site of Special Scientific Interest (SSSI).[2,21]

A report for the Colonial Office, prepared by Stewart Lockhart in anticipation of the extension of British control to include the New Territories in 1898, included observations on the absence of extensive forests.[22] Some cultivation of *Pinus massoniana* on lower hills for firewood was noted, as was the maintenance of 'clumps of well-grown trees' near many villages (*fung shui* woods) and the remnants of forest patches in sheltered ravines. The colonial government instigated an extensive afforestation programme in the mid-1870s, with a million or more trees (primarily *Pinus massoniana*) planted annually during the following decade; this scheme initially focused on Hong Kong Island, but was subsequently extended to include the New Territories after 1900.[2,23] It is estimated that by 1938 as much as 70% of Hong Kong Island consisted of plantations resulting from this afforestation scheme.[2]

The Second World War had a devastating impact on Hong Kong's tree flora. The Japanese invasion of China disrupted fuel supply lines, forcing the inhabitants of Hong Kong to rely on local forests for firewood. The subsequent invasion of the territory in December 1941 further exacerbated the problem, resulting in large-scale deforestation; this continued even after liberation until adequate fuel supply lines could be re-established in 1946.[2,23] Post-war aerial photographs reveal the extent of the devastation, with a return to barren, treeless landscapes that would have been easily recognisable to Hinds and Seemann a century earlier. As in the nineteenth century, most of the forests that remained were either those associated with villages (*fung shui* woods) or those in isolated ravines, protected by their inaccessibility.

The post-war history of Hong Kong has been characterised by rapid population growth and urbanisation, with infrastructure development, land reclamation and associated loss of agriculture.[24] The colonial government initiated a new afforestation programme in 1953, again focusing primarily on the planting of *Pinus massoniana*, supplemented by exotic species including *Eucalyptus robusta*, *Casuarina equisetifolia*, *Lophostemon confertus* and *Melaleuca cajuputi*.[23,25] Although this post-war reafforestation was of a much smaller scale than that between the 1870s and 1930s, it is estimated that by 1959 forests covered 4,800 ha (*c.* 4.5%) of land in the territory.[23]

The mid-1960s saw a significant shift in environmental policy in Hong Kong, with the publication of two important government reports. P. A. Daley[26] emphasised the importance of managing forests for the prevention and amelioration of soil erosion and to meet the cultural and recreational needs of the community. L. M. & M. H. Talbot[27] furthermore identified the need for establishing legally protected areas in Hong Kong; this proposal ultimately led to the creation of Country Parks in 1976,[28] which now protect over 40% of land in the territory.[23]

As a result of the government's reafforestation scheme, native Chinese red pines (*Pinus massoniana*) again became a dominant component of Hong Kong's forests by the end of the 1970s. Concern was raised in 1978, however, by the death of many *Pinus massoniana* trees following the wilting and browning of needles.[23] It was determined that this was due to 'pine wilt disease', resulting from infestation by pinewood nematodes (*Bursaphelenchus xylophilus*) that are spread between trees by longicorn beetles (Cerambycidae).[3] The effects of the disease were limited within its natural range in North America, but caused widespread destruction of local *Pinus massoniana* populations, dramatically reducing the extent of forested land in Hong Kong. The extent of the ecological damage caused by pine wilt disease led to the recognition of the importance of species diversity in afforestation programmes.[23] Despite the devastation caused by the disease, the loss of large tracts of *Pinus massoniana* provided opportunities for the natural succession of native trees such as *Schefflera heptaphylla* and *Machilus* species.[23]

The tree flora of Hong Kong can be classified into four categories based on their ecology and species composition[29] (listed here with exemplar species illustrated in this book): (1) riverine forests (including *Syzygium jambos*); (2) lowland forests, 300–400 m above sea level (including *Acronychia pedunculata*, *Schefflera heptaphylla*, *Sterculia lanceolata*, and species in the families Euphorbiaceae, Moraceae, Myrtaceae and Sapotaceae); (3) lower montane forests, 300–800 m above sea level (including *Machilus* species, *Schima superba*, and species in the families Fagaceae, Lauraceae and Theaceae); and (4) montane forests, 700–1,000 m above sea level (including *Illicium angustisepalum* and species in the families Fagaceae, Hamamelidaceae, Magnoliaceae and Theaceae).

In addition to these essentially natural forest types, many traditionally protected *fung shui* woods have been maintained in close proximity to villages in accordance with the Chinese system of geomancy. In terms of their species composition, *fung shui* woods typically resemble other lowland forests but have often been enriched by the cultivation of species that are useful to the neighbouring villages, such as *Aquilaria sinensis* (the source of incense), *Cinnamomum camphora* (camphor wood), *Dimocarpus longan* (longans), *Litchi chinensis* (lychees) and *Syzygium jambos* (rose apples).[23,30,31] Many of these *fung shui* woods are of considerable age (those in Shing Mun Country Park are reputedly 400 years old), but based on their species composition are nevertheless regarded as secondary forests (*i.e.*, resulting from natural succession following anthropogenic activities, including agriculture).[2] True primary forests no longer exist in Hong Kong.

—— ∽ ——

The Evolutionary Diversity of Trees

> It is interesting to contemplate an entangled bank, clothed with many plants of many kinds, with birds singing on the bushes, with various insects flitting about, and with worms crawling through the damp earth, and reflect that these elaborately constructed forms, so different from each other, and dependent on each other in so complex a manner, have all been produced by laws acting around us. … from so simple a beginning endless forms most beautiful and most wonderful have been, and are being, evolved.
>
> Charles Darwin, *On the Origin of Species* (1859)

Woody plants are those that exhibit secondary growth (lateral thickening) of stems and roots, enabling the broad trunk characteristic of trees. This feature appeared early in the evolutionary history of plants and most contemporary woody plants share the same developmental process for achieving secondary growth, irrespective of whether they are gymnosperms (seed-bearing plants in which the seeds are not enclosed in a carpel, which first appeared in the Late Carboniferous, *c.* 320–290 million years ago[32]), or whether they are angiosperms (flowering plants, which appeared much later, in the Early Cretaceous, *c.* 130 million years ago[33]). The monocotyledons (commonly known as 'monocots') are an important evolutionary lineage of flowering plants that evolved from a common ancestor that had lost the ability to achieve secondary growth; several monocots have nevertheless independently regained the capacity for secondary growth, but in these cases the developmental mechanism causing secondary growth is very different from that observed in other woody plants.[33] The most prominent woody monocots in the Hong Kong flora—palm trees (family Arecaceae) and screw pines (family Pandanaceae)—lack 'true' secondary growth, however, and achieve their thickened stems by extensive lateral growth immediately behind the apical growing point.

Pines (in the genus *Pinus* of the family Pinaceae), which are good examples of gymnosperm trees, bear male and female reproductive cones rather than flowers. The male cones are comparatively small and are spirally clustered around small branches, and consist of a series of fertile leaf-like structures ('sporophylls') that bear 'sporangia' that produce and enclose the developing pollen; the pollen grains in turn develop an internal structure that produces the male gamete (sperm). The sporophylls of the male cone separate slightly at maturity, with the pollen dispersed from the sporangia by wind. The female cones are much larger and more complex structures in which the ovules (which contain the egg cells) are unprotected and borne directly on the upper surface of large scales. The apex of the ovule is exposed at maturity and forms a small droplet of sticky fluid that traps the wind-borne pollen. The pollen grain subsequently germinates, forming a pollen tube that conveys the sperm to the egg, inside the ovule. After fertilisation, the ovule develops to form the seed, which is itself wind-dispersed.

Of the 19 cone-bearing (gymnosperm) trees listed in *Flora of Hong Kong*,[34] only three are regarded as native or truly naturalised:[1] *Amentotaxus argotaenia* (in the family Cephalotaxaceae), *Keteleeria fortunei* (Pinaceae), and *Pinus massoniana* (Pinaceae). Of these three, only *Pinus massoniana* is included in this book, although we have also illustrated the closely related and widely planted exotic *Pinus elliottii* and *Podocarpus macrophyllus*.

The vast majority of native and naturalised tree species in Hong Kong are flowering plants (angiosperms). Flowers consist of four main classes of floral organ, arranged in the following sequence from the outer margin of the flower to the centre: (1) the sepals (collectively known as the calyx), which in most cases perform a protective function, enclosing the flower bud during development; (2) the petals (collectively known as the corolla), which in most cases assist as a visual cue to attract pollinators; (3) the stamens (collectively known as the androecium), which produce the pollen grains, which in turn produce the male gametes (sperm); and (4) the carpels (collectively known as the gynoecium), which contain the ovules, which in turn produce the female gamete (the egg). This enumeration of the four classes of floral organs is greatly simplified, however, as there is an enormous diversity of flower morphology associated with different pollination and reproductive systems: not all floral organs are present in all species; not all organs retain their ancestral function, sometimes adopting alternative roles; and not all floral organs remain as separate structures, often becoming fused, either with other organs of the same type (*e.g.*, petal fused with petal, or carpel with carpel), or else with different types of organ (*e.g.*, petal fused with stamen, or stamen with carpel).

Pollen is transferred from the stamen of one flower to the carpel of another by a variety of mechanisms (although this sometimes occurs within a flower, resulting in self-pollination). In the majority of angiosperm species, pollen is transferred by animals, although there is an enormous diversity of different co-evolutionary specialisations, including those associated with insects and vertebrates.[35,36] The ancestral pollination system in angiosperms is likely to have been insects (particularly beetles).[37] Beetle-pollinated flowers are often comparatively robust (especially if the pollinating beetles are large), have whitish, yellow, brown or red petals, and often have a strong fruity or spicy scent; the food reward offered to the beetles is typically pollen. Most fly-pollinated flowers are pale coloured with a fruity scent, with nectar and/or pollen as a reward. Bee pollination is very widespread in angiosperms, accounting for c. 65% of all species; bee-pollinated flowers are often white, yellow or blue (often with floral 'guides' and asymmetric petals to encourage the bees to enter the flower in a specific direction) and sweet-scented, with nectar and/or pollen again provided as a food reward. Moth- and butterfly-pollinated flowers are generally intensely perfumed, with nectar presented as a food reward at the base of a long floral tube that the moth or butterfly probes with its elongated proboscis. Butterfly-pollinated species are often colourful (red, blue, yellow or sometimes white) and are receptive during the day, whereas moth-pollinated flowers are pale and receptive at dusk or at night, when the moths are active. Most vertebrate pollinators are either birds or bats. Bird-pollinated flowers are typically brightly coloured (especially red), but generally lack a distinct scent; they produce copious nectar to meet the significant energy demands of the bird. Bats are active at dusk and night, and flowers that are adapted to bat pollination are consequently pale-coloured; the bats are attracted by strong, musky, fruity or foetid scents and are rewarded by large quantities of nectar and pollen.

This diversity of biotic pollination systems has resulted from extensive co-evolution, in which floral specialisations are adaptations for different pollinators that respond to specific cues and seek specific rewards. There is also evidence for multiple parallel evolutionary shifts from biotic to abiotic pollination systems in angiosperms, with c. 20% of all species pollinated by wind. Wind pollination inevitably results in greater pollen wastage, although this is compensated by a reduced energy investment by the plant: the flowers are highly reduced (generally lacking petals, for example), and there is no need to provide energy-rich food rewards such as nectar. Wind-pollinated flowers are typically unisexual, and are therefore better able to avoid self-pollination. Many tree species are wind-pollinated, typically growing in dense populations and often in relatively open, more exposed habitats.

Unlike gymnosperms, in which the ovules are unprotected, the ovules of angiosperms are enclosed within carpels (the female organ of the flower). Successful pollination results in fertilisation of the egg cell in the ovule, and the ovule develops into the seed. In angiosperms, the carpel that surrounds the ovule develops into the fruit: although they are commonly referred to as 'flowering plants' since they are defined by the possession of flowers, angiosperms could equally well be referred to as 'fruiting plants.'

Fruits are extremely diverse morphologically,[38] in part because of the complex way in which they develop from the flower: although it is the fertilised carpel that forms the main component of the fruit, this is often supplemented by other floral tissues (including the calyx and corolla), and adjacent flowers in an inflorescence sometimes coalesce to form a single 'multiple' fruit. In the same way that flowers have diversified in response to co-evolution with pollinators, fruits have diversified in response to different seed dispersal agents, including wind, water, and animals.[39] Seeds that are wind-dispersed typically either have broad wings (*e.g., Pinus massoniana* and *Polyspora axillaris*) or else are attached to a mass of cotton-like fibres (*e.g., Bombax ceiba*) that enable the seeds to remain airborne for longer. Seeds that are water-dispersed generally have internal air chambers to increase buoyancy (*e.g., Hibiscus tiliaceus*). Many fruits are eaten by animals, with the seeds either discarded during chewing, regurgitated after swallowing, or defecated after passing through the animal's gut. Such animal-dispersed fruits are typically fleshy, with different characteristics (size, colouration and nutritive content) associated with adaptations to different animal groups. Birds are the most common seed dispersers in Hong Kong,[40] with most fruits limited in size by the beak gape of the bird. Other important seed dispersers in Hong Kong include fruit bats, civets, macaques, and possibly also ferret badgers and muntjacs (barking deer).[2]

—— ℰℑ ——

Botanical Nomenclature

"Let us go for a walk tomorrow morning," she said to him. "I want you to teach me the Latin names of our wild flowers and their characteristics."

"What do you want with Latin names?" Bazarov asked.

"System is needed in everything," she replied.

Ivan Turgenev, *Fathers and Sons* (1862)

The application of scientific names is strictly controlled by internationally accepted codes of nomenclature; in the case of plants, this is the *International Code of Nomenclature for Algae, Fungi, and Plants* (ICN).[41] The ICN recognises a hierarchy of nested taxonomic units, in which groups of related species are aggregated into larger units known as genera, and related genera are similarly aggregated into families.

Scientific names are Latin (or 'latinised,' in which words derived from other languages are treated as though they are of Latin origin), and hence by convention are italicised. Species names ('binomials') consist of two component words: the genus name, written with a capital first letter; followed by the 'specific epithet,' entirely in lower case. Citations of species names are commonly followed by the name of the taxonomist who first described the species and coined the name: the name *Acacia confusa* Merr., for example, was first described by the American taxonomist Elmer D. Merrill (1876–1956). Subsequent research by other taxonomists sometimes results in the transfer of a species to a different genus. Since the generic name forms the first part of the species name, a new binomial is created by combining the initial specific epithet with the newly associated generic name. In these cases, the name of the taxonomist first describing the species is cited in parenthesis, followed by the name of the person effecting the transfer: the name *Alangium chinense* (Lour.) Harms, for example, was first described in the genus *Stylidium* by João de Loureiro in 1790 as *Stylidium chinense* Lour., before being transferred to the genus *Alangium* by Hermann A. T. Harms in 1897.

One of the basic principles of the ICN is that each taxonomic unit, at whatever rank in the hierarchy, should only be assigned one name, which is unique to it. The transfer of species between genera described in the previous paragraph inevitably gives rise to duplicate names ('synonyms') for the same species. Synonymy can also result from taxonomic errors, however, in which a duplicate name is published because the taxonomist was not aware that the species had already been described; this was the case, for example, when Augustin Pyramus de Candolle published the name '*Bombax malabaricum* DC.' in 1824 for a species that had previously been described by Carolus Linnaeus in 1753 as '*Bombax ceiba* L.' Since the ICN requires the recognition of only one name for each species, the earliest validly published name is adopted (the 'principle of priority'); in the example given, the name *Bombax ceiba* is adopted as the accepted name, and the name *Bombax malabaricum* is treated as a synonym.

Although species are generally reproductively isolated from other species, preventing genetic mixing and the consequent breakdown of taxonomic distinctions, hybridisation can sometimes occur. Hybridisation is known to have occurred, for example, between *Bauhinia purpurea* and *Bauhinia variegata*; the resultant hybrid is cited by separating the specific epithets of the two parental species by the symbol '×' (in this case as '*Bauhinia purpurea* × *variegata*'). If the hybrid is able to reproduce but is reproductively isolated from its parental species, it is typically recognised as a distinct species in its own right and assigned a species binomial. *Bauhinia purpurea* × *variegata* is sterile, however, and since it is reliant on artificial propagation is recognised here as a 'cultivar' (a plant variety produced and maintained in cultivation). Cultivar names are not italicised, and are cited in inverted commas (in this example, *Bauhinia* 'Blakeana').

Plant families (consisting of closely related genera, derived from a common evolutionary ancestor) are assigned names that almost invariably possess the Latin ending '-aceae'—in the case of *Bauhinia*, for example, the family name is the Fabaceae. These family names are provided in the text accompanying each of the species illustrated, following the citation of the English and Chinese vernacular names.

——— ❧ ———

The Art of Botanical Illustration

Seeing no way to preserve this marvelous form, I attempted an exact drawing of it, whereby I deepened my insight into the fundamental concept of metamorphosis.

Johann Wolfgang von Goethe, *The Metamorphosis of Plants* (1790)

Effective communication is of fundamental importance in comparative biology: this communication is achieved in part by the clarity of written descriptions but also by the accuracy of visual representations. Although photography plays a crucial role in conveying information on plant structure in contemporary botanical research it cannot supplant the highest-quality illustrations, in which the artist can highlight structures of particular importance in ways that are not possible using photography. Botanical illustration transcends the divide between purely scientific representation, in which accurate interpretation of morphology is paramount, and the expressive arts, in which representation of form can extend beyond the constraints of reality as a means of conveying the artist's stylistic intent. Botanical art of the highest calibre is based not only on accurate observations of plant structure, but also achieves heightened artistry in terms of overall composition and rendition: the visual representation of the plant should serve equally well as a research tool on a scientist's bench as it should as a framed picture on a gallery wall.

Some of the earliest surviving botanical illustrations date from the sixth century in a Byzantine manuscript known as the *Codex Vindobonensis*.[42] These early illustrations were invariably reproduced in 'herbals'—manuscripts that provided guidance for apothecaries on the medicinal use of plants.[43] Prior to the invention of printing, these herbals were copied by hand with the inevitable consequence that the botanical illustrations often became reduced to mere caricatures, with stylised and often exaggerated features. The invention of printing with moveable type in Germany in the mid-fifteenth century revitalised the publication of herbals, enabling the creation of multiple copies of each book without the inevitable artifacts arising from manual copying. The invention of printing furthermore coincided with a broader intellectual renaissance in Europe, in which classical sources of knowledge were rediscovered and new scientific methods developed that were in turn instrumental in heralding the modern age of scientific investigation. The artistic genius of the Renaissance artists Leonardo da Vinci (1452–1519) and Albrecht Dürer (1471–1528) demonstrated the remarkable fidelity with which plant structure could be represented, and this greatly influenced subsequent artists, including those who worked for the botanists and herbalists Otto Brunfels (1488–1534), Leonhart Fuchs (1501–66) and Pier Andrea Mattioli (1501–77).

The art of botanical illustration received considerable impetus in the sixteenth and seventeenth centuries from the European royal houses, which were keen to amass collections of paintings of native and exotic plants. These artistic advances were also promoted by the scientific revolution that was integral to the seventeenth- and eighteenth-century 'Age of Enlightenment' and the influx of new and exotic plants that were being brought to European nations from their expanding empires. European enthusiasm for the cultivation of garden exotics grew hand-in-hand with the publication of increasingly affordable and accessible botanical illustrations. *Curtis's Botanical Magazine* began publication in 1787 with hand-coloured copper engravings of selected species. The journal is still published today and is the world's longest-running botanical magazine[44] (although it has appeared under several different names, including *The Kew Magazine* between 1984 and 1994). Botanical illustration has a long history and continues to play a key role in plant science research.

Plant Portraits

——— ❧ ———

Acacia auriculiformis Cunn. ex Benth.

———— ❧ ————

Acacia auriculiformis (Ear-pod wattle or Ear-leaved Acacia; 耳果相思 or 耳葉相思; Fabaceae)[45] is native to northern Australia and New Guinea but has been extensively planted in Hong Kong. Many species of *Acacia* have 'bipinnate' leaves consisting of leaflets ('pinnae') that are themselves divided into higher-order leaflets. In some acacias, however, including *Acacia auriculiformis*, the foliage is highly reduced, with only a single pinnate leaf evident in seedlings; as the plant matures the photosynthetic function is transferred instead to expanded and leaf-like petioles known as 'phyllodes.'[46,47] It appears that the evolutionary replacement of leaf laminas with phyllodes has been driven by the selective advantage of greater resistance to periods of water stress and high irradiance.[47,48] The phyllodes are associated with small extra-floral nectaries[49] that secrete a sugary fluid that attracts ants, possibly benefitting the plant by deterring potential herbivores.

As with *Acacia confusa* (*q.v.*), which is also widely planted locally, the flowers of *Acacia auriculiformis* are very small and aggregated into clusters (inflorescences). The two species are easily distinguished by the shape of these inflorescences, however: those of *Acacia auriculiformis* are elongated (3.5–8 cm long), whereas those of *Acacia confusa* are spherical (6–10 mm in diameter). The two species also differ in the shape of the fruit pods, which are straight and elongated in *Acacia confusa* but distinctly curved and 'ear-like' in *Acacia auriculiformis*, giving rise to its English vernacular names.

The fruit pods of *Acacia* dry out and split open at maturity to expose the seeds. The seeds of *Acacia auriculiformis* possess a two-lobed yellow, orange or red 'aril,' which is a fleshy elaboration of the outer layer of the seed coat. The aril functions as a food reward to attract animals, thereby promoting seed dispersal. The seeds of *Acacia* species that have arils are typically bird dispersed,[50] and it is not surprising that *Acacia auriculiformis* seeds have been observed locally to be eaten by birds even though the species in not native to Hong Kong.[51] Although *Acacia auriculiformis* is widely planted in Hong Kong and is able to set fruit and successfully disperse its seeds, there is no evidence that it has become naturalised.[51]

Plate 1. *Acacia auriculiformis.* (A) Flowering branch, showing elongated inflorescences composed of highly reduced flowers. (B) Dissected flower. (C) Fruit. (D) Seeds with two-lobed aril.

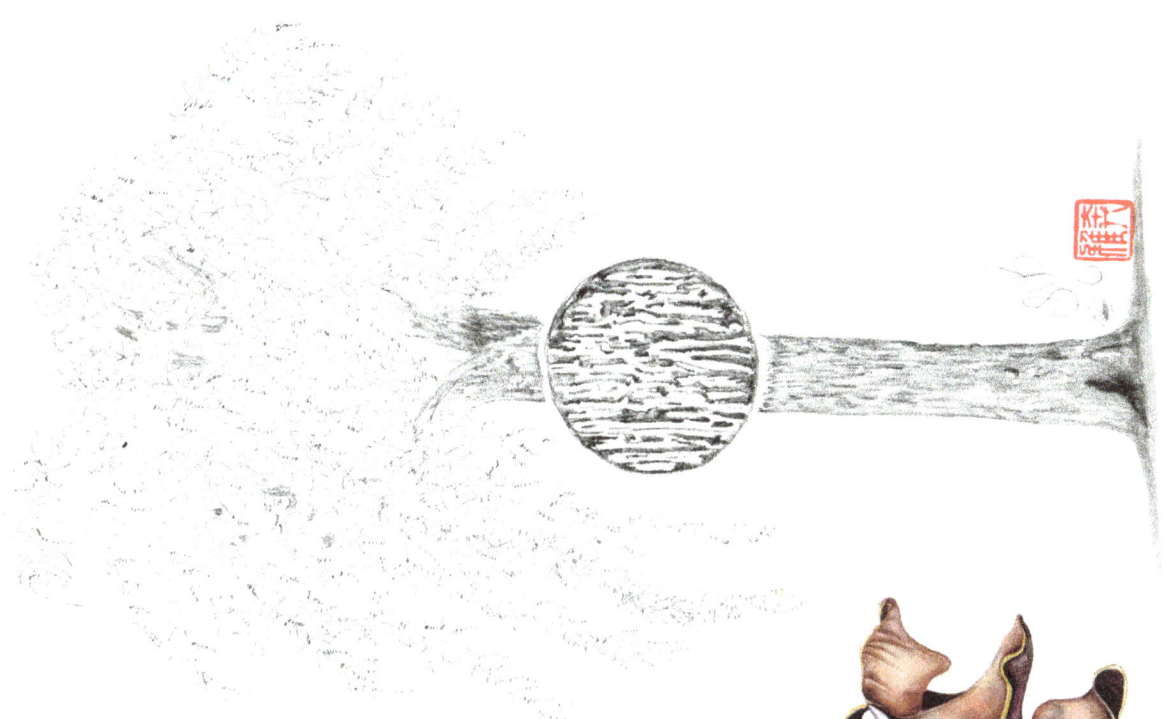

Acacia confusa Merr.

(= *Acacia richii* auct. non A. Gray)

—— ℘ ——

Acacia confusa (Taiwan Acacia; 台灣相思; Fabaceae)[45] is native to Taiwan and the Philippines but has been extensively cultivated in Hong Kong and southern China. As with *Acacia auriculiformis* (*q.v.*), the leaves of *Acacia confusa* are reduced to expanded and leaf-like petioles known as 'phyllodes.' These phyllodes (6–10 cm long) have between three and five parallel veins, and, as in *Acacia auriculiformis*, are associated with small extra-floral nectaries.[49]

The flowers are very small and are aggregated into spherical inflorescences (6–10 mm in diameter); the petals in each flower are fused into a short tube (approximately 2 mm long) with numerous stamens that extend beyond the apex of the petal tube.[45] The flowers are commonly visited by bees.[52]

Acacia confusa grows well in degraded areas and hence is valuable for preventing continued soil erosion.[53] It is also comparatively fire-resistant and has accordingly been widely cultivated in Hong Kong as a fire break in plantations;[25] anthropogenic hill fires are common in the region, often associated with traditional 'grave sweeping' festivals. *Acacia confusa* is of low ecological value compared to native species, however, and local tree species are now generally preferentially planted as a more valuable ecological resource.

Acacia confusa is known to excrete toxic biochemicals into the soil following the rotting of fallen leaves. The chemicals released into the soil in this way act to inhibit or suppress the growth of other plants.[54] This phenomenon—known as 'allelopathy'—is beneficial to the tree itself as it reduces below-ground competition for nutrients and water, but is an undesirable characteristic in a plantation species as it hinders natural regeneration and biodiversity restoration.

Plate 2. *Acacia confusa.* Flowering branch, showing leaf-like phyllodes and spherical inflorescences of highly reduced flowers.

Acer sino-oblongum F. P. Metcalf

(= *Acer lanceolatum* auct. non Molliard; *Acer oblongum* auct. non Wall. ex DC.)

——— ☙ ———

Acer sino-oblongum (South China maple; 濱海槭, 華南飛蛾樹 or 劍葉槭; Sapindaceae)[55] is locally common[1] and is one of three native maple species in Hong Kong. Unlike most maple species, which have palmately lobed leaves, the leaf laminas of *Acer sino-oblongum* are entire and elliptical or slightly oblong. The flowers are rather inconspicuous, greenish-yellow, and clustered in terminal inflorescences. Each small flower has five sepals and five petals (*c.* 4 mm long) and is either male (with eight stamens and lacking carpels) or structurally bisexual but functionally female (with two fused carpels and sterile stamens). This separation of floral sex prevents self-pollination within the flower and hence promotes the genetic diversity of the seeds produced. The flowers contain a glandular disc that produces a nectar food reward for pollinating insects; studies of other Asian species have highlighted the important role of bees as pollinators.[56]

The pistil in functionally female flowers is derived from two fused carpels, each of which develops a lateral wing after fertilization; these wings become very prominent in mature fruits (B in the accompanying plate). The fruits are described as 'schizocarpic,' reflecting the fact that they split into two separate dispersal units at maturity: each unit is derived from a single carpel with the seed borne on one side and with a solitary wing extending outwards on the opposing side. The centre of gravity is therefore located at the side with the seed, and the lateral wing rotates around the seed as the fruit falls from the tree branch; this slows the descent of the fruit and allows greater opportunities for cross-winds to disperse the seed further afield.

Acer species have traditionally been classified in the family Aceraceae—the approach adopted, for example, in the *Flora of Hong Kong*.[55] Recent reconstructions of the evolutionary tree of maples and related taxonomic groups, however, consistently show that *Acer* is nested within the family Sapindaceae;[57,58] although some taxonomists have dissented,[59] the current consensus favours this broader circumscription of the Sapindaceae.[14]

Plate 3. *Acer sino-oblongum.* (A) Flowering branch, showing terminal inflorescences composed of small greenish-yellow flowers. (B) Fruiting branch, with schizocarpic fruits.

A

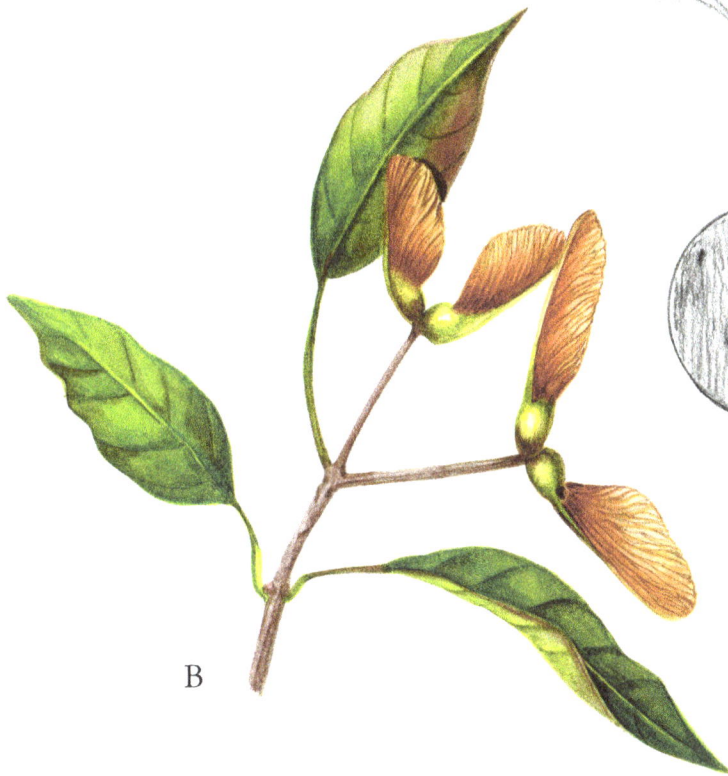

B

Acronychia pedunculata (L.) Miq.
(= *Acronychia laurifolia* Blume; *Cyminosma pedunculata* DC.)

——— ✿ ———

Acronychia pedunculata (Acronychia; 山油柑 or 降真香; Rutaceae)[60] is very common in lowland secondary forests in the region, comprising small trees growing to 15 m in height. It belongs to the same plant family as the commercially cultivated citrus fruits, and its leaves are characterised by small oil glands that are visible if the leaves are held up against the light, producing a strong citrus-like odour if crushed.

The flowers are 12–15 mm in diameter, with four narrow yellowish-white petals, eight stamens arranged in two whorls, and four carpels that are fused to form the pistil. The species is primarily pollinated by nymphalid butterflies but are also visited by a diverse range of other insects, including bees, wasps, calliphorid flies, papilionid butterflies and beetles.[61]

Acronychia pedunculata fruits are small and rounded (10–15 mm in diameter), translucent-yellow, and very fragrant when crushed. Each fruit contains four seeds that are enclosed within the hard, protective inner layer of the fruit wall. The fruits are known to be eaten by birds and civets (and possibly also fruit bats), with the seeds dispersed after passing through the animal's gut.[62,63] Chemical analysis of the sugar content of the fruits reveals that they are sucrose-dominated, in contrast with most bird-dispersed fruits in which glucose and fructose dominate.[63] Although this may represent an adaptation favouring a greater diversity of seed dispersers, including mammals such as civets, it is also significant that *Acronychia pedunculata* fruits are eaten by bulbuls (*Pycnonotus* species), which are unusual amongst birds in possessing intestinal sucrase enzymes that are capable of digesting sucrose.[63]

Plate 4. *Acronychia pedunculata.* (A) Flowering branch, showing clusters of small yellow-white flowers. (B) Flower. (C) Fertilised flower, after the petals and stamens have fallen. (D, E) Fruits.

B

C

D

E

Adenanthera microsperma Teijsm. & Binn.

(= *Adenanthera pavonina* auct. non L.; *Adenanthera pavonina* var. *microsperma*
(Teijsm. & Binn.) I. C. Nielsen)

———— ❦ ————

Adenanthera microsperma (Red sandalwood; 海紅豆 or 孔雀豆; Fabaceae)[45] has a rather restricted distribution in Hong Kong, although it is locally common in *fung shui* woods near villages.[1] The trees grow to *c*. 20 m and possess characteristic compound leaves composed of lateral 'pinnae' that are subdivided into smaller, alternately arranged leaflets; each of these leaflets (2.5–3.5 cm in length) are rounded in outline.

The flowers are small, white or yellow, and borne in long inflorescences up to *c*. 15 cm. Each flower comprises a tiny five-toothed fused calyx, five small petals (2.5–3 mm long), 10 stamens and a single carpel. The stamens bear small anther glands that presumably function by providing a food source to reward pollinators,[64,65] although surprisingly little is known about the pollination ecology of *Adenanthera* species.

The fruit pod is large, up to 20 cm in length, and splits into two valves as it dries, with each valve twisting in opposite directions. As the pod opens, the bright red glossy seeds are exposed, suspended from the pale brown inner surface of the pod. These seeds superficially resemble fleshy berries and it has been suggested that the red colouration has evolved to mimic berries and deceive birds that normally eat fleshy fruits:[66] experiments have shown that *Adenanthera* seeds are refused by birds that specialise in eating dry fruits and seeds but are accepted by birds that eat fleshy fruits; in the latter case, the seeds are defecated intact. It therefore seems likely that frugivorous birds are effective seed dispersers of *Adenanthera microsperma*, without the necessity for the tree to invest energy-expensive resources in providing a food reward for the birds.

Plate 5. *Adenanthera microsperma*. (A) Flowering branch, with elongated inflorescences composed of small white or yellow flowers. (B) Leaf. (C) Flower. (D) Mature fruit, with separated valves twisting in opposite directions, exposing the bright red seeds.

Adina pilulifera (Lam.) Franch. ex Drake

(= *Adina globiflora* Salisb.)

—— ⁊ ——

Adina pilulifera (Chinese buttonbush or Pilular Adina; 水團花; Rubiaceae)[67] is a small tree (growing to *c.* 5 m) that is very common beside streams in Hong Kong. As with most members of the species-rich family Rubiaceae, *Adina pilulifera* is characterised by a very regular opposite-paired arrangement of branches and leaves.

The flowers are very small, with the five white petals fused into a corolla tube that is 3–5 mm long and 2–3 mm in diameter. Numerous flowers are aggregated into distinctive spherical inflorescences, 8–12 mm in diameter. The pollen-receiving stigma is borne at the apex of an elongated style that protrudes from the mouth of each corolla tube, giving the inflorescence its distinctive 'spiked' appearance. The anthers are fused to the apex of the corolla tube so that pollinators will inadvertently collect pollen as they probe into the flower to collect nectar, whilst brushing past the stigmas and depositing pollen grains as they navigate across the surface of the inflorescence.

Before the flowers open at sexual maturity, the inflorescences superficially resemble the aggregate fruits of the Strawberry tree, *Myrica rubra* (Myricaceae, *q.v.*); this similarity, together with the riverine habitat, has led to the Chinese vernacular name for *Adina pilulifera*, which translates as 'Aquatic strawberry tree.'

The seeds are small and winged, and are likely to be wind dispersed, although given the close proximity of the trees to streams water dispersal also seems feasible; both dispersal mechanisms have been suggested for other species in the same genus.[68]

Plate 6. *Adina pilulifera.* (A) Flowering branch, showing spherical inflorescences formed from numerous small flowers. (B) Flower, with fused corolla tube and protruding stigma and style.

A

B

Adinandra millettii (Hook. & Arn.) Benth. & Hook. f. ex Hance

——— ❧ ———

Adinandra millettii (Millett's Adinandra; 黃瑞木; Pentaphylacaceae)[69] is a small shrubby tree, growing to 10 m. It is locally common in secondary forests and shrubland, especially in granite areas.[1] The flowers are borne at the base of the leaves and are generally pendent; each flower is bisexual, with five sepals (7–8 mm long), five white petals (*c.* 9 mm long, and protruding slightly beyond the top of the calyx), *c.* 25 stamens and a single fused pistil. The sepals persist even after the flower has been fertilised, and remain clasping the base of the black, globose fruit. These fruits are likely to be bird-dispersed,[63] and it is significant that although the fruits are rich in glucose and fructose sugars, there is no detectable sucrose:[63] this is typical of bird-dispersed fruits, since birds generally lack the sucrase enzyme necessary to digest this type of sugar.

Adinandra millettii was historically classified in the plant family Theaceae—the approach adopted, for example, in the *Flora of Hong Kong*[69] and *Flora of China*.[70] Recent analysis of DNA sequence data, however, has enabled a better-resolved reconstruction of evolutionary relationships within the group,[71] and this has revealed that the species previously classified in the Theaceae represent two distinct evolutionary lineages that are not each other's closest relatives. Several species, including *Adinandra millettii*, have consequently been segregated into the family Pentaphylacaceae.[13]

The leaves of *Adinandra millettii* (Yang Tong, 杨桐) were traditionally used as a dye in the manufacture of coloured glutinous rice balls served during the Ching Ming (清明節) grave-sweeping festival.[72]

Plate 7. *Adinandra millettii.* (A) Flowering branch, with pendent axillary flowers. (B) Single flower, with white petals. (C) Black pendent fruits, with persistent sepals.

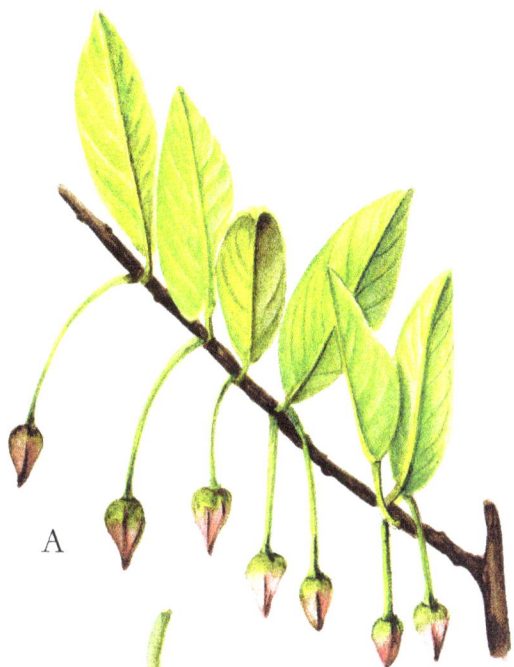

A

B

C

Alangium chinense (Lour.) Harms

(= *Marlea begoniaefolia* Roxb.; *Stylidium chinense* Lour.)

— ❧ —

Alangium chinense (Chinese Alangium; 八角楓; Alangiaceae)[73] is a common tree in southern China and Southeast Asia. Seasonality of growth is very clear, with the current year's green shoots developing from axillary buds on older, brown branches; the history of this episodic growth pattern is evident as a series of conspicuous annual scars on the branches.[74] The leaves are very variable in shape, ranging from ovate with entire margins (as shown in the accompanying plate) to distinctly lobed.

The elongated flower buds open in a very characteristic way, with the petals (6–8 per flower) curling back and changing colour from cream to yellow as they age. This movement of the petals exposes the stamens, which consist of elongated pollen-bearing anthers that are borne on short, hairy filaments. The stamens surround the central pistil, which is tipped with a 2–4-lobed stigma. The flowers are pollinated by bees (*Apis cerana*) that are attracted by nectar.[61] The fruits are small, black, single-seeded 'drupes' (fleshy berry-like fruits in which the seed is surrounded by a hard protective layer), and are consumed by birds, which are effective seed dispersers for the species.[62]

Alangium chinense is widely used in Chinese medicine and is considered one of the '50 fundamental herbs'. The shoots and bark of the species are used to treat a diverse range of medical conditions including snake bites, rheumatism, circulatory problems, and is also used as a contraceptive.[75] The species has consequently been the focus of considerable pharmacological research.[76]

Plate 8. *Alangium chinense.* (A) Flowering branch, showing a young green shoot growing from a leaf axil on a branch from the previous year. (B) An open flower, with recurved petals. (C) Pistil. (D) Stamen, with an elongated anther borne on a short, hairy filament. (E) Fruit, with persistent calyx.

A

B

C

D

E

Aleurites moluccana (L.) Willd.

—— ❧ ——

Aleurites moluccana (Candlenut tree; 石栗; Euphorbiaceae)[77] is one of Southeast Asia's most important multi-purpose trees; the great range of uses for the tree encouraged its early introduction across Southeast Asia, where it is often grown in home gardens. The seeds are oil-rich and have traditionally been used for lighting (hence its English vernacular name) and as a food,[78] although the seeds are poisonous[79] and require cooking to break down the toxins. The seeds have also been used as a soap substitute and hair conditioner, and are now important commercial products in the cosmetics industry.[78] Various other plant organs, particularly the leaves, bark, flowers, fruits and seeds, have been used in local medicine for treating a wide variety of ailments.[78] The wood is extensively used locally: although it is comparatively weak and is not resistant to fungal decay or insect attack, it is easily worked by hand.[78] The species is highly tolerant of air pollution and hence is very extensively planted along roadsides in Hong Kong, although the lack of strength in the wood means that trees often suffer typhoon damage.[80]

As with many species in the family Euphorbiaceae (*e.g.*, *Endospermum chinense*, *Macaranga tanarius*, *Mallotus paniculatus*, *Sapium discolor* and *Vernicia montana*, *q.v.*), the leaves are adorned with a pair of extra-floral nectaries at the junction between the leaf lamina and the petiole.[49] These nectaries presumably function by encouraging ant colonies, which are likely to deter potential herbivores.

Aleurites moluccana trees are bisexual but bear separate male and female flowers in the same inflorescence: the male flowers are smaller but more numerous than the female flowers. This separation of male and female flowers (known as 'monoecy') is an evolutionary adaptation to reduce the possibility of self-fertilisation, thereby increasing the genetic diversity of the seeds produced. The fruits are fleshy, with a hard inner layer surrounding two seeds.

Plate 9. *Aleurites moluccana.* (A) Flowering branch, showing inflorescence composed of numerous small flowers. (B) Flower. (C) Dissected male flower. (D) Fruit.

Anneslea fragrans Wall. var. *hainanensis* Kobuski

(= *Anneslea hainanensis* (Kobuski) Hu)

—— ∽ ——

Anneslea fragrans var. *hainanensis* (Hainan Anneslea; 茶梨; Pentaphylacaceae)[69] is very rare in Hong Kong, only having been recorded from Ma On Shan.[1] It is a medium-sized evergreen tree, growing to *c.* 15 m in height, with alternately arranged leathery leaves that are crowded towards the ends of branches. The flowers are bisexual, with five reddish sepals (*c.* 10–15 mm long) that are basally conjoined and markedly constricted in the middle; these sepals are persistent, forming a distinctive crown over the apex of the fruit. Each flower also has five red or white petals (*c.* 6–8 mm long) that are basally fused, *c.* 40 stamens, and a pistil formed from the fusion of three carpels.

Anneslea fragrans flowers are structurally similar to those of *Pentaphylax euryoides* (*q.v.*), which is also classified in the Pentaphylacaceae, although the flowers of the latter only have five stamens. *Anneslea fragrans* flowers furthermore have 'semi-inferior' ovaries that are located partly below the point of attachment of the perianth; in contrast, *Pentaphylax euryoides* has 'superior' ovaries that are located above the point of attachment of the perianth. Although the position of the ovary might superficially appear trivial, the evolutionary origin of inferior or semi-inferior ovaries endows a major selective advantage by protecting the ovary (and hence the ovules with their eggs) from floral visitors and potential herbivores:[81] inferior and semi-inferior ovaries are known to have evolved independently in many evolutionary lineages, in which it often represents a key evolutionary innovation.[33]

Another curiosity of floral structure in *Anneslea fragrans* is the position of the petals relative to the sepals: in most flowers (including *Pentaphylax euryoides*), the petals alternate with the sepals; in *Anneslea fragrans*, however, the petals develop opposite—and hence lie directly on the inside of—the sepals.[82]

Plate 10. *Anneslea fragrans* var. *hainanensis*. (A) Flowering branch, bearing a cluster of bright red flowers. (B) Flower, with petals located opposite the sepals. (C) Fruit, with its crown of persistent sepals.

A

B

C

Annona squamosa L.

———— ☙ ————

Annona squamosa (Sugar-apple or Sweetsop; 番荔枝; Annonaceae)[83] is a small tree (growing to *c.* 4 m) that is widely cultivated in tropical regions for its edible fruit. The species originated in the West Indies but appears to have been cultivated in Asia for many centuries: there is strong linguistic and cultural evidence that it was introduced to the Philippines and India by Spanish and Portugese colonists, respectively; and it appears to have been cultivated in Java prior to the arrival of the Dutch in the seventeenth century.[84]

Annona squamosa belongs to an early-divergent evolutionary lineage of flowering plants and possesses many characteristics that are recognised as ancestral. Unlike most early-divergent flowering plants (such as *Illicium angustisepalum* and *Magnolia championii, q.v.*), however, Annonaceae species—including *Annona squamosa*—have structurally very distinct sepals and petals: the Annonaceae resembles most of the derived flowering plant lineages in this respect, but it is likely that this similarity has evolved independently.[85]

Annona squamosa flowers are bisexual, with sexual function extended over a two-day period: the carpels are receptive during the first day, with the stamens releasing pollen the following day. This temporal separation of female and male reproductive function (a phenomenon known as 'protogyny') effectively avoids within-flower self-pollination and hence helps promote genetic diversity in the seeds produced.[86] The flowers are known to be visited by small nitidulid beetles, which are attracted by the fruity floral scent and are effective pollinators.[87]

Each flower has numerous unfused and spirally arranged carpels—another ancestral characteristic widespread in early-divergent flowering plants. After fertilisation, however, these separate carpels coalesce during fruit development;[88] this is presumably an adaptation enhancing frugivory (and hence seed dispersal) by larger mammals such as primates.

Plate 11. *Annona squamosa.* (A) Fruiting branch. (B) Flower. (C) Cross-section through fruit, showing fused fruit segments. (D) Seeds.

A

B

C

D

Aporosa dioica (Roxb.) Müll. Arg.

(= *Alnus dioica* Roxb.; *Aporosa chinensis* (Champ. & Benth.) Merr.;
Aporosa frutescens auct. non Blume; *Aporosa leptostachya* Benth.)

——— ❧ ———

Aporosa dioica (Aporosa; 銀柴; Euphorbiaceae)[77] has had a complex nomenclatural history. William Roxburgh—the first taxonomist to describe the species in 1832—was initially misled by the appearance of the catkin-like clusters of highly reduced flowers into believing that the tree was a type of alder: Roxburgh accordingly give it the name *Alnus dioica*,[89] although the genus *Alnus* is now classified in the distantly related family Betulaceae.[13] The species was subsequently correctly classified in the family Euphorbiaceae as a species of *Aporosa* (although an alternative spelling, *Aporusa*, has often been used),[90] under a variety of different specific epithets.

The trees are 'dioecious,' meaning that individuals bear flowers of only one sex. The catkins borne on male individuals are noticeably longer than those on females, and are visited by *Apis cerana* bees, which collect pollen grains.[61] These bees do not visit female catkins, however, and therefore cannot enable pollination. It seems likely that the species is actually wind-pollinated:[61] the highly reduced floral morphology and separation of sexes are common features of wind-pollinated species. In a similar way, bees are known to visit male and not female flowers of the related species *Mallotus paniculatus* (*q.v.*), which is also dioecious with highly reduced flowers.

The fruits are small capsules that split open irregularly at maturity, exposing the brightly coloured seeds. Reports from Hong Kong reveal that the fruits are eaten by birds,[62] which presumably act as the primary seed dispersal agent; reports from other regions, however, indicate that gibbons may also be important agents for dispersing the seeds.[91]

The stems of *Aporosa dioica* are commonly attacked by gall midges (Cedidomyiidae), which cause the formation of conspicuous swollen galls that are characteristic of the species.[92]

Plate 12. *Aporosa dioica*. (A) Flowering branch with elongated male inflorescences. (B) Male flowers. (C) Female flower. (D) Fruiting branch with bright red seeds.

B

A

C

D

Aquilaria sinensis (Lour.) Spreng.

(= *Aquilaria grandiflora* Benth.)

———— ♋ ————

Aquilaria sinensis (Incense tree; 土沉香, 牙香樹 or 白木香; Thymelaeaceae)[93] is relatively common in Hong Kong, especially in *fung shui* woods.[30] It is unclear whether the species is truly native, but if not, it has been fully naturalised for a long time.[1] The trees are the source of a highly fragrant, resinous wood (known commercially as 'agarwood') that is used in the manufacture of incense and as a source of Chinese medicine (沉香).[94] The high concentration of resin in the agarwood appears to develop in response to fungal infection, with uninfected trees lacking the valuable resin.[95] Mature individuals of *Aquilaria sinensis* are often indiscriminately felled in Hong Kong in the hope of retrieving commercially valuable agarwood.[96] The species is now regarded as vulnerable to extinction[97] and is legally protected in China.[98]

Hong Kong was historically an important hub in the commercial export of *Aquilaria sinensis* incense to other parts of Asia, with most of the incense exported via Shek Pai Wan (Aberdeen). This harbour became known as 'Heong Kong' (which translates as 'Incense Harbour'), a name that was subsequently misapplied (as 'Hong Kong') to the entire territory.[99]

The yellowish-green flowers of *Aquilaria sinensis* are borne in small inflorescences. Each flower is bisexual, with 10 tiny petals at the top of a tube formed from five fused sepals.[93] The fruits are obovoid capsules that split into two valves at maturity; the seeds (only one or two per fruit) hang suspended from the fruit by long, silky threads.[94] The seed dispersal mechanism is complex: the seeds are unlikely to be bird-dispersed since there is little obvious nutritive value other than a small fleshy basal appendage (the 'caruncle'). Other species in the genus have been shown to have only limited dispersal capability and hence have been inferred to be gravity-dispersed.[100] An intriguing observation suggests, however, that the seeds are probably dispersed by vespid wasps: wasps have been observed to land on the suspended seed, breaking the silk-like thread, and then carrying the seed away (80 m on average).[101] It is likely that the vespid wasps are deceived by the resemblance of the seed to a caterpillar suspended on its thread.

Plate 13. *Aquilaria sinensis.* (A) Flowering branch. (B) Single flower, showing five sepal lobes. (C) Mature dehisced fruit, with seeds suspended on threads.

Archidendron lucidum (Benth.) I. C. Nielsen

(= *Abarema lucida* (Benth.) Kosterm.; *Pithecellobium lucidum* Benth.)

———— ✌ ————

Archidendron lucidum (Chinese Apea ear-ring; 亮葉猴耳環; Fabaceae)[45] is an ecological pioneer, common in shrubland and near the margins of secondary forests in Hong Kong.[1] The success of the species during the early stages of ecological succession is promoted in part by the occurrence of small root nodules that contain colonies of symbiotic filamentous bacteria;[102] these bacteria are able to convert free atmospheric nitrogen into nitrogen-rich compounds that can be utilized by the tree. *Archidendron lucidum* does not generally persist into later successional, closed-canopy forests, however.[103]

The trees bear rounded clusters of 10–20 small flowers, each of which has fused white petals (4–5 mm long) and numerous long stamens that protrude from the flower, giving the inflorescence a feathery appearance.

After fertilisation, the flowers develop into large fruit pods (15–20 cm long, and 2–3 cm wide) similar to those observed in *Adenanthera microsperma* (*q.v.*), which belongs to the same plant family. The pod splits open along two sutures are maturity, with the two valves twisting in opposite directions, coiling to form a distinctive spiral. Unlike *Adenanthera microsperma*, however, the fruit pod of *Archidendron lucidum* is bright red at maturity, with blue-black seeds that hang suspended from the fruit wall. The contrasting colours of the seeds and the fruit wall are likely to be features that have evolved to promote seed dispersal by birds.[68,103] *Archidendron lucidum* possibly adopts a similar dispersal mechanism to that observed in *Adenanthera microsperma*, in which fruit-eating birds are deceived into swallowing seeds that resemble small berries.[68]

Plate 14. *Archidendron lucidum.* (A) Flowering branch, with rounded inflorescences composed of small, cream-coloured flowers. (B) Single flower. (C) Mature, open fruit pod, with blue-black seeds suspended from the bright red fruit wall.

A

B

C

Ardisia quinquegona Blume
(= *Ardisia pauciflora* Heyne ex Roxb.)

———— ℰℛ ————

Ardisia quinquegona (Asiatic Ardisia; 羅傘樹; Myrsinaceae)[104] is a very common understorey shrub or small tree species, growing to 5 m in height.[1] As the specific epithet implies, the flowers have organs arranged in multiples of five: five fused white or pink petals; five stamens, each of which is located on the inside of a petal and fused to it; and a pistil derived from five carpels.

The single-seeded fruits are 5–7 mm in diameter, fleshy and slightly five-angled. They are presumably eaten by birds,[63] which disperse the seeds: the fruit flesh has been shown to be rich in glucose and sucrose sugars[63] as is typical for bird-dispersed fruits; and each fruit (a 'drupe') has a tough inner layer surrounding the seed that provides protection as the seed passes through the bird's digestive tract.

Many plant species have evolved symbiotic associations with bacteria that can fix atmospheric nitrogen that would not otherwise be available to the host plant: this is generally evident as bacterial colonies within root nodules (observed, for example, in *Archidendron lucidum* and *Casuarina equisetifolia, q.v.*). Many *Ardisia* species (including the Hong Kong native *Ardisia crenata*, although not *Ardisia quinquegona*) have an unusual variant of this symbiosis: instead of root nodules, these *Ardisia* species possess bacteria-containing nodules along the leaf margins.[105] Symbiotic bacteria are also present in an apical chamber enclosed by young leaves at the shoot tip, where they are supported by mucilage secreted by hairs on the upper leaf surface;[106] newly developing leaves are inoculated with these bacteria via pores along the leaf margins, resulting in the formation of new marginal leaf nodules.[107] The complexity of the symbiosis even extends to inter-generational inheritance due to the incorporation of bacterial cells in the developing flower bud, with the embryo developing within the bacteria-filled mucilage and incorporating the bacteria into its primordial shoot tip.[108]

Plate 15. *Ardisia quinquegona*. (A) Flowering branch, showing loose inflorescences of pinkish-white flowers. (B) Flowers. (C) Fruiting branch, with five-angled fleshy drupes.

Artocarpus hypargyreus Hance ex Benth.
(= *Ficus laceratifolia* auct. non Lévl. & Van.)

—— ❧ ——

Artocarpus hypargyreus (Silver-back Artocarpus or Sweet Artocarpus; 白桂木; Moraceae)[109] is locally common in Hong Kong[1] and cultivated in *fung shui* woods.[30] It belongs to the same plant family as fig trees (*e.g., Ficus variolosa, q.v.*), a family that is characterised by very highly reduced flowers that are tightly aggregated into condensed inflorescences, subsequently developing into fleshy infructescences ('syncarps'). Unlike figs, however, in which the inflorescences are completely invaginated hollow structures, *Artocarpus* species possess rounded inflorescences with the flowers borne externally. Individual flowers are unisexual, with inflorescences composed of flowers of a single sex.

The fleshy fruit syncarps (3–4 cm in diameter) are sweet and particularly rich in sucrose and glucose.[63] In Hong Kong, the fruits are removed directly from the trees by macaques (*Macaca* species), which eat the fruit pulp and spit the seeds out; the distances over which seeds are dispersed is increased by the short-term storage of food in the macaques' cheek pouches.[2] It is clear that other modes of seed dispersal must exist, however, since macaques have not been recorded in Hong Kong for most of the past 150 years: the fruits that fall to the ground in forests not inhabited by macaques are reported to be eaten by civets, which are also likely to act as seed dispersers.[62,110]

Artocarpus hypargyreus is closely related to two economically important species, *Artocarpus altilis* (Breadfruit) and *Artocarpus heterophyllus* (Jackfruit), both of which are cultivated for their fleshy fruits.[111] Breadfruit was identified in the eighteenth century as a potential staple food crop for slaves in the West Indies. The infamous story of the first attempt to collect living trees from Tahiti has become part of our cultural and historical heritage: the story of Lieutenant William Bligh's lengthy voyage from England to Tahiti (1787–88) aboard *H. M. S. Bounty*, the mutiny led by Fletcher Christian in April 1789, Bligh's tortuous 3,600-mile open-boat voyage to Timor, and the subsequent recriminations and court-martials have become literary and cinematic legends.[112,113]

Plate 16. *Artocarpus hypargyreus.* (A) Flowering branch, showing tightly compact inflorescences of highly reduced flowers. (B) Mature fruits (syncarps).

A

B

Aucuba chinensis Benth.

———— ∾ ————

Aucuba chinensis (Chinese Aucuba; 桃葉珊瑚; Garryaceae)[114] is a shrubby species growing to 6 m (rarely as small trees to 12 m), locally common in Hong Kong on Tai Mo Shan, Sunset Peak and at Wong Nei Chong.[1] *Aucuba* species are very variable in habit and leaf shape and taxonomic opinion on the number of species in the genus is accordingly diverse, ranging from only one highly variable species (*Aucuba japonica*) up to 12 species.[115] *Aucuba chinensis* is one of three long-recognised species, however: it was first described in 1861 by George Bentham in his *Flora Hongkongensis*[8] based on specimens collected on Hong Kong Island by the colonial surgeon W. A. Harland (1818–58).

Aucuba chinensis bears separate male and female flowers that are strikingly 'tetramerous' (with organs in groups of four): in addition to four tiny sepals and four conspicuous purplish-red petals (3–4 mm long), the male flowers have four stamens that alternate with the petals. The two sexes of flower are noticeably different in overall shape: the carpel in female flowers is located below the point of attachment of the perianth (often described as having an 'inferior' ovary) and hence the female flower appears elongated below the petals. This combination of tetramerous flowers with an inferior ovary led many taxonomists to classify *Aucuba* in the family Cornaceae (as adopted, for example, in the *Flora of Hong Kong*[114]) or else to separate it in its own family, the Aucubaceae, allied to the Cornaceae.[116] Recent phylogenetic studies based on DNA sequence data, however, have indicated a close affinity with the North American genus *Garrya*,[117] and both are now classified in the family Garryaceae, close to the Eucommiaceae.[13]

The fruits are bright red 'drupes' (fleshy fruits with a tough inner fruit wall surrounding the single seed), 14–18 × 8–12 mm. Although the dispersal agent was not identified in a study of Hong Kong populations,[62] the fruits are known to form part of the diet of macaques elsewhere.[118]

Plate 17. *Aucuba chinensis*. (A) Inflorescence of male flowers. (B) Male flower, showing four stamens alternating with the purplish-red petals. (C) Female flower. (D) Fruiting branch, showing red drupes.

Bauhinia purpurea × *variegata* 'Blakeana'
(= *Bauhinia blakeana* Dunn)

—— ∾ ——

Bauhinia purpurea × *variegata* 'Blakeana' (Hong Kong orchid tree; 洋紫荊; Fabaceae)[119] is an artificially propagated hybrid derived from interbreeding between two naturally occurring species. It was first found growing in the grounds of an abandoned house near Mount Davis, Hong Kong Island, by a French missionary in the late nineteenth century and was transplanted to the grounds of the Pokfulam Sanatorium (now Bethanie), then run by the Missions Étrangères de Paris,[120] and from there introduced to the Hong Kong Botanic Gardens. The latter individual (which survived being blown over by a severe typhoon in 1906) was used to propagate new trees vegetatively after 1914,[121] and it is likely that all individuals in Hong Kong and elsewhere are descended from this single tree.

The tree was named *Bauhinia blakeana* in 1908 in honour of Sir Henry Blake,[122] Governor of Hong Kong between 1898 and 1903. Recent research has confirmed its hybrid status, resulting from interbreeding between *Bauhinia purpurea* and *Bauhinia variegata*.[123,124] Although hybridisation plays a key role in plant speciation if sterility can be overcome, *Bauhinia blakeana* is reliant on artificial propagation, generally by grafting onto the root-stocks of other *Bauhinia* species. As it is unable to reproduce itself, it is inappropriate to regard it as a distinct species and it therefore does not warrant a species binomial: the tree was accordingly assigned the cultivar name 'Blakeana' in 2005.[123]

Bauhinia purpurea × *variegata* 'Blakeana' was adopted as the floral emblem of Hong Kong in 1965,[80] and now appears, often in a stylised form, on local banknotes, coins and the SAR flag. As a sterile hybrid, this is arguably an inauspicious symbol for a city built on mixed Chinese and British heritage.

Plate 18. *Bauhinia purpurea* × *variegata* 'Blakeana'. (A) Flowering branch, showing characteristic bilateral symmetry of the flower, five stamens and single carpel. (B) Leaf. (C) Curved carpel, with elongated stalk. (D) Curved stamen.

Bombax ceiba L.
(= *Bombax malabaricum* DC.)

———— ❧ ————

Bombax ceiba (Cotton tree; 木棉; Malvaceae)[125] is widespread in tropical Southeast Asia. Although not indigenous to Hong Kong it has been extensively cultivated locally as an ornamental tree along roadsides, lending its name to 'Cotton Tree Drive' in the Central District of Hong Kong.

The species is deciduous and has a very distinctive architecture, with horizontally spreading branches and characteristic conical spines on the main trunk. The flowers are conspicuous not only because of their large size (approximately 10 cm in diameter) and deep red colour, but also because they are borne in March and April, before the flush of that season's leaves. The flowers are pollinated by a diverse range of birds and bats[126,127] that are attracted by the copious nectar, although some bird species are known to puncture a hole in the side of the unopened flower bud to steal nectar.

The structure and arrangement of the stamens in the flower is unusual, with three distinct types evident.[128] The outermost stamens are basally fused into five clusters (each with 6–31 stamens) that alternate with the five petals. The inner stamens form a ring around the fused carpels and are of two lengths: five longer stamens, each with two large pollen sacs (E in Plate 19); and ten shorter stamens, each with a single pollen sac (F in Plate 19).

The fruit capsule matures in May, bursting open to reveal a mass of cotton-like fibres resembling that of Kapok (*Ceiba pentandra*) that assists with the wind dispersal of the small seeds (Plate 20). Unlike true cotton, however, *Bombax* fibres are shorter and are not flattened and so do not twist and cannot be woven to make fabric; the fibres are therefore of only limited value, and can only be used as stuffing for cushions, *etc.*

Plate 19. *Bombax ceiba* (fruits shown in Plate 20). (A) Flowering branch, showing flower buds and mature flower at different developmental stages. The monochrome drawing of the entire tree illustrates the characteristic branching pattern and trunk spines. (B) Flower (longitudinal section), showing stamens of differing lengths. (C, D) Floral receptacle (entire and longitudinal section). (E) Longer stamen, showing double pollen sacs. (F) Shorter stamens, showing single pollen sac.

Plate 20. *Bombax ceiba* (flowers shown in Plate 19). (A) Unopen fruit. (B) Open fruit, revealing mass of cotton-like fibres. (C) Small seeds amongst the fibres.

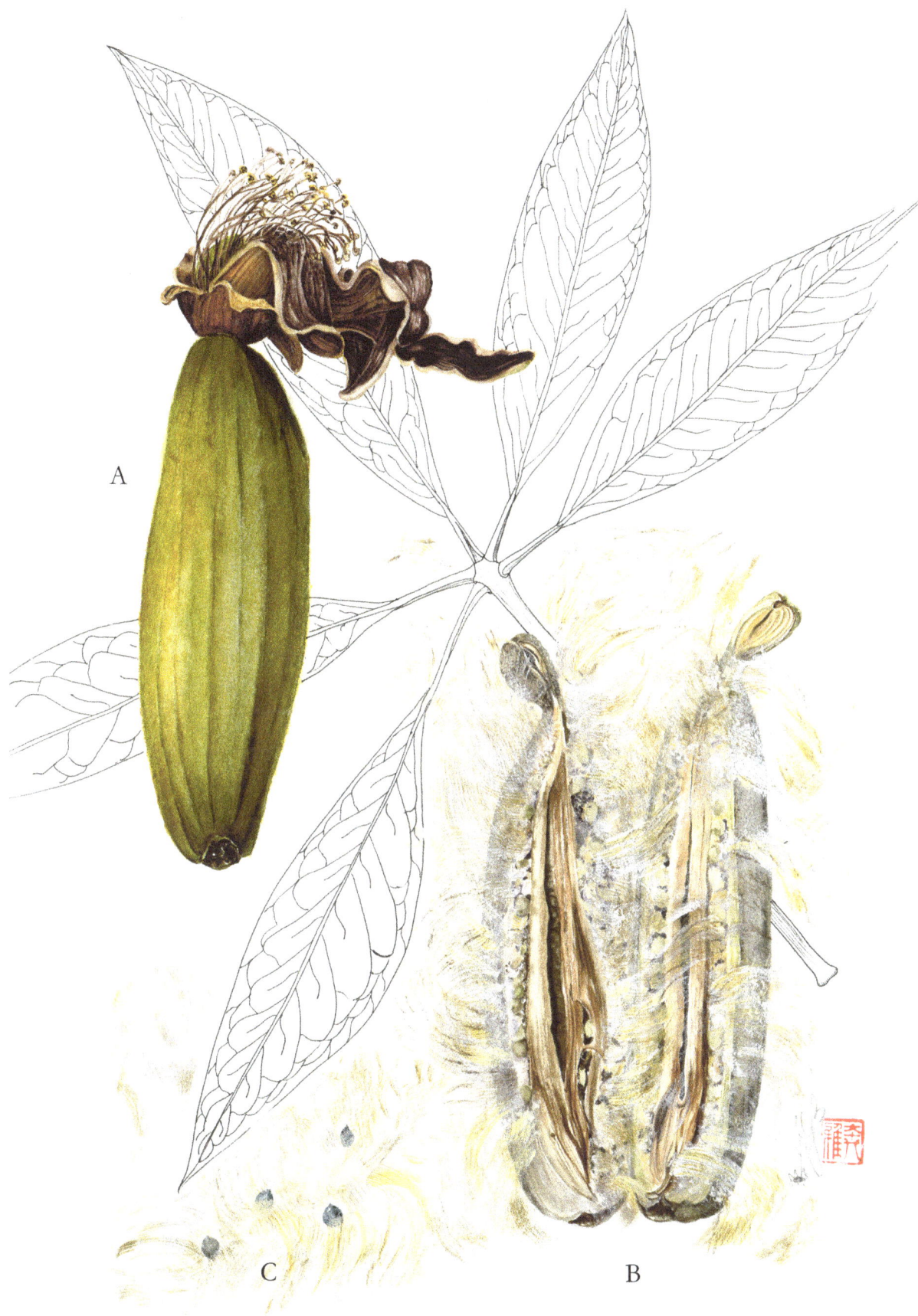

Bruguiera gymnorhiza (L.) Savigny
(= *Rhizophora gymnorhiza* L.)

—— ❧ ——

Bruguiera gymnorhiza (Many-petaled mangrove; 木欖; Rhizophoraceae)[129] is one of only eight plant species in Hong Kong that are truly adapted to survive in mangrove habitats (three of which—see also *Heritiera littoralis* and *Kandelia obovata*—are illustrated in this book). Mangroves are intertidal ecosystems in which organisms experience major fluctuations in salinity associated with freshwater discharge and periodic tidal flushing; as a consequence, the habitat is extremely challenging for plant survival, with associated risks of high temperature, desiccation, oxygen deprivation and shifting substrates. Many of the morphological peculiarities of mangrove trees are adaptations to survival in these difficult conditions, including 'prop' roots that help support the tree on soil that is unstable because of tidal inundations, roots with 'knees' that project above the soil surface to assist with aeration in oxygen-poor soil, and the evolution of 'vivipary,' in which seeds germinate whilst still attached to the maternal plant[130,131] and hence enable rapid seedling establishment in unstable conditions.

Bruguiera gymnorhiza flowers are pendent, with 11–13 deep-red sepals that are basally fused into a tube up to 2 cm in length, and an equivalent number of petals, each adorned with three or four apical bristles. The flowers are likely to be bird-pollinated, with Common white-eyes (*Zosterops japonicus*) drinking nectar with the aid of a brush-like adaptation to their tongue.[2]

The seeds germinate precociously whilst still attached to the maternal plant—a phenomenon known as 'vivipary.' Germination begins soon after fertilisation, leading to the formation of an elongated 'hypocotyl' (the lower part of the immature shoot, below the 'cotyledons' or first-formed leaves) that superficially resembles a fruit. These seedlings are very similar to those of the closely related species *Kandelia obovata* (*q.v.*), which belongs to the same plant family; in both species the precocious development of the seedlings enables them to become established very quickly.

Plate 21. *Bruguiera gymnorhiza.* (A) Fertile branch, showing flowers (above) and viviparous seedlings (below), surrounded by the persistent calyx composed of fused red sepals. (B) Petal, with apical bristles. (C) Base of trunk, showing distinctive 'prop' roots.

A

B

C

Callicarpa nudiflora Hook. & Arn.

(= *Callicarpa reevesii* Wall. ex Schauer)

———— ✺ ————

Callicarpa nudiflora (Callicarpa; 裸花紫珠; Lamiaceae)[132] is a pioneer shrub or small tree, growing to 4 m (rarely to 7 m), and common in shrubland.[1] Although the flowers are small, they are an attractive pink or purple colour and are aggregated into large rounded inflorescences, 8–13 cm in diameter. Each flower has a cup-shaped calyx derived from the fusion of four sepals, and petals that are similarly fused, forming a tube *c.* 2 mm long. The style and stamens extend well beyond the apex of the corolla tube, giving the inflorescence a less compact appearance. Surprisingly little has been published on the pollination ecology of the genus, although the flowers are likely to attract butterflies,[133] which presumably transfer pollen between the anthers and stigmas as they probe into the flowers for nectar.

The fruits are small purple 'drupes' (berry-like fruits with a tough inner fruit layer surrounding the seeds). These characteristics are typical of bird-dispersed fruits:[62] the colour of the fruit skin is within the bird's visual perception range; the size (*c.* 2 mm in diameter) is within the limits of the bird's beak gape; and the tough inner fruit wall protects the seed as it passes through the bird's gut. Analysis of the sugar content of a related *Callicarpa* species has revealed that the fruits are rich in glucose and fructose, as is typical for bird-dispersed fruits.[63]

Callicarpa species, including *Callicarpa nudiflora*, are very important in traditional Chinese medicine, and a considerable body of literature exists on their phytochemistry and pharmacology. Studies indicate that *Callicarpa* species have anti-inflammatory, immunological, haemostatic, neuroprotective and analgesic properties, amongst others.[134]

Plate 22. *Callicarpa nudiflora.* (A) Flowering branch, showing large inflorescence composed of numerous small purple-pink flowers with stamens that extend beyond the top of the corolla tube. (B) Single flower. (C, D) Clusters of purple fruits.

Camellia crapnelliana Tutch.

———— ❧ ————

Camellia crapnelliana (Crapnell's Camellia; 紅皮糙果茶; Theaceae)[69] possesses a conspicuous orange-red bark that easily distinguishes it from other local camellias. William J. Tutcher discovered a single individual of the species growing on the southern slope of Mount Parker, Hong Kong Island, in 1903, which he subsequently formally described and named in 1905.[135] Very few natural populations survive in Hong Kong: in addition to the original collection locality, it is only known from Mau Ping in Sai Kung (which has been recognised as a Site of Special Scientific Interest because of the occurrence of this species).[136] *Camellia crapnelliana* is now legally protected both locally and nationally,[136] with seedlings cultivated *ex situ* for reintroduction into the wild.[98,137]

The flowers are comparatively large, 6–10 cm in diameter, with petals that are 3.5–6.5 cm long,[69] although there are other *Camellia* species in Hong Kong with larger flowers. Each flower has 6–8 showy white petals, numerous stamens with long golden-yellow filaments, and a compound pistil derived from the fusion of 3–5 carpels.

The fruits are large (5–10 cm in diameter) and dry. The seeds are likely to be dispersed by rodents, which store them for later consumption: surplus seeds and those that are subsequently forgotten (or those stored by a rodent that dies) have the opportunity to germinate.[2] This type of seed dispersal, known as scatter hoarding, has been demonstrated for other *Camellia* species[138] and is also observed in many Fagaceae species (*e.g.*, *Castanopsis fissa*, *q.v.*), which have similarly large, dry fruits.[2] Despite the large number of tree species in Hong Kong that bear fruits that appear to be adapted for scatter-hoarding dispersal, rodents capable of effecting this dispersal no longer survive in Hong Kong.[139] Species showing this dispersal mechanism, including *Camellia crapnelliana*, are accordingly often relatively rare and are noticeably absent from secondary forests.[110]

Plate 23. *Camellia crapnelliana*. (A) Flowering branch, with large, showy flower. (B) Large, dry fruit.

A

B

Camellia hongkongensis Seem.

---— ✃ ——

Camellia hongkongensis (Hong Kong Camellia; 香港茶; Theaceae)[69] is the only native camellia species in Hong Kong with red flowers: all others are white (*e.g.*, *Camellia crapnelliana, q.v.*), although several of the locally cultivated species are pink or red. The flowers have six or seven petals (3.5–4 cm long) that are slightly fused basally. Each flower has numerous yellow stamens (*c.* 3 cm long) that are fused to the base of the petals. The pistil has three or four styles that are not fused, distinguishing the species from the locally cultivated *Camellia japonica* which also has red flowers, but which has fused styles.

Camellia hongkongensis was first discovered near Tourane in Cochinchina (now Da Nang, Vietnam) in 1837 by the French botanist Charles Gaudichaud-Beaupré during his final circumnavigation of the globe.[140] The species was subsequently collected in Hong Kong in 1849 by Lieutenant-Colonel John Eyre, although only three individuals were discovered locally.[140] Early botanists believed that it represented a wild form of the widely cultivated *Camellia japonica*,[141] but it was subsequently recognised as a new species by Berthold Seemann in 1859.[140] Attempts at incorporating the species in afforestation programmes during the late nineteenth century were of only limited success,[25] and the species remains very rare locally, only known from a few localities on Hong Kong Island and Lamma.[136] *Camellia hongkongensis* is now legally protected under the Forestry Regulations of the Hong Kong Government's Forests and Countryside Ordinance and is actively propagated for reintroduction into the wild.[98]

Camellia hongkongensis fruits are hard 'capsules,' *c.* 3 cm in diameter.[69] As with other *Camellia* species,[138] the seeds are likely to be dispersed through 'scatter hoarding' in which rodents store the seeds for later consumption (see discussion under *Camellia crapnelliana*). The ecological loss of suitable rodent dispersal agents from Hong Kong presumably explains the comparative rarity of species such as *Camellia hongkongensis*.[139]

Plate 24. *Camellia hongkongensis.* (A) Flowering branch, with large, showy flowers. (B) Dry fruit.

A

B

Castanopsis fissa (Champ. ex Benth.) Rehder & E. H. Wilson

(= *Quercus fissa* Champ. ex Benth.)

——— ✑ ———

Castanopsis fissa (Chestnut oak; 黧蒴錐; Fagaceae)[142] is a common local tree, closely related to the oak genus, *Quercus*, in which it was originally classified. The flowers are unisexual, highly reduced and are borne in spike-like inflorescences that are 8–15 cm long; clusters of these spikes are aggregated into large 'panicles'. Each panicle consists of mostly male spikes below a terminal female spike, although spikes with a mix of male and female flowers sometimes occur between the unisexual spikes. The female flowers are enclosed by a 'cupule' formed from fused branches,[143] covered externally by concentric rings of tubercles (but lacking the spines that are characteristic of many other species in the genus).[142] Although *Castanopsis* species can have up to three female flowers within the cupule, this has been reduced to only one in *Castanopsis fissa*.[144]

The flowers are primarily visited by *Apis cerana* bees, although other common visitors include calliphorid flies and lycaenid butterflies.[61] After fertilisation, the cupules expand, reaching up to 1.5 cm in diameter and splitting into 2–4 irregular segments.[142] A single nut is borne within each fruit cupule.

The fruits are probably dispersed by scatter-hoarding rodents.[110,145] The fauna of Hong Kong no longer includes such dispersal agents,[139] however, and as a result *Castanopsis fissa* and related species are potentially vulnerable in human-dominated landscapes such as Hong Kong. *Castanopsis fissa* has been extensively planted as part of the territory's reafforestation policy, however, with nursery-grown saplings often planted into established *Pinus massoniana* plantations.[25]

Plate 25. *Castanopsis fissa.* (A) Flowering branch, showing large panicles of highly reduced flowers. (B) Fruits within their cupules.

Casuarina equisetifolia L.

—— ❧ ——

Casuarina equisetifolia (Horsetail tree or She-oak; 木麻黃 or 牛尾松; Casuarinaceae)[146] grows to 35 m and is characterised by slender pendent branches that have a distinctive 'jointed' appearance, with whorls of 6–8 very highly reduced leaves encircling the branch at each node. The evolutionary reduction of the leaves reduces water loss and is presumably an adaptation enabling survival in dry and saline conditions; photosynthetic function has shifted from the leaves to the slender branches, which are green due to the presence of chlorophyll. The 'jointed' appearance of the branches superficially resembles the stem of the herbaceous fern ally *Equisetum* (also known by the English vernacular name Horsetail);[147] this resemblance has led to the species epithet '*equisetifolia*' (literally meaning '*Equisetum*-leaved') and vernacular name 'Horsetail tree.'

Casuarina equisetifolia is native to Australia but has been extensively planted locally as part of the Government's reafforestation programme[25] as it is fast-growing and is able to tolerate comparatively infertile soils. These characteristics are largely due to a symbiotic association that is maintained in small nodules on the tree's roots with colonies of nitrogen-fixing filamentous bacteria, which provide the tree with important nitrogen-rich nutrients in the form of ammonia.[148] *Casuarina equisetifolia* is accordingly well adapted to infertile sandy soils, and its ability to survive saline conditions—including periodic inundations by sea water—has led to its widespread use for the stabilisation of sand dunes and control of coastal erosion.[80]

The highly reduced flowers are unisexual and borne concurrently on the same tree: the male flowers are clustered in small spikes (1–4 cm long) at the end of branches, whereas the female flowers are aggregated in rounded lateral clusters. Pollination is achieved by wind[149] and results in the formation of fruiting structures that superficially resemble miniature pine cones (B, C in Plate 26) but which consist of pairs of hard woody bracts that enclose individual fruits that are derived from each female flower. The bracts separate to release the single-seeded fruits (D in Plate 26) that are winged (known as 'samaras') and wind-dispersed.[68]

Plate 26. *Casuarina equisetifolia.* (A) Fruiting branch, showing slender green branches with whorls of very highly reduced leaves. (B) Immature green multiple fruit prior to the release of the samaras. (C) Brown multiple fruit after separation of the woody bracts. (D) Winged fruits (samaras).

A

B

C

D

Celtis sinensis Pers.

(= *Celtis tetrandra* Roxb. subsp. *sinensis* (Pers.) Y. C. Tang)

—— ❧ ——

Celtis sinensis (Chinese hackberry or Chinese nettle tree; 朴樹; Cannabaceae)[150] is a common native tree that has also been extensively planted as an ornamental in parks and along roadsides. Although the species is slow growing, it is versatile in its soil requirements, develops a deep root system and has strong wood, and is therefore very resistant to typhoon damage and is often long-lived.[80] The species is also tolerant of air pollution and is therefore an ideal shade tree for planting along roadsides in Hong Kong.[80]

Although many tree species in Hong Kong are evergreen (or are deciduous but with a very brief period in which they are leafless), *Celtis sinensis* has an extended period of more than a month each year without leaves.[2]

Celtis sinensis was historically classified in the elm family, Ulmaceae, which is characterised by small flowers that lack petals. The flowers of *Celtis sinensis* have four or five sepals, with an equal number of stamens located immediately above each sepal. Although some flowers lack female reproductive organs and are therefore functionally male, other flowers borne on the same individual have a compound pistil (derived from the fusion of two carpels) and are therefore functionally bisexual. The flowers are pollinated by *Apis cerana* bees.[61]

The fruits are small red-brown 'drupes' (berry-like fruits with a hard inner fruit wall), 5–7 mm in diameter,[150] with a single seed, *c.* 4.5 mm in diameter. The fruits are eaten by birds, which are effective seed dispersers.[62] Interestingly, the number of fruits borne by individual trees greatly exceeds the number eaten by birds: the excess fruit remains on the tree and becomes dry, although further research is necessary to provide a scientific explanation for this.

Plate 27. *Celtis sinensis.* (A) Fruiting branch, showing slightly immature fruits. (B) Abaxial surface of the leaf. (C) Bisexual flower, with four stamens and a single compound pistil. (D) Male flower, structurally similar to the female flower, but lacking the pistil.

Cerbera manghas L.
(= *Cerbera odollam* auct. non Gaertn.)

——— ❦ ———

Cerbera manghas (Cerbera or Sea mango; 海杧果; Apocynaceae)[151] is a common littoral species and is widespread in tropical Asia. The leaves and flowers are characteristically crowded at the ends of branches. As is typical for members of the family Apocynaceae, the plants exude sticky white latex when broken; this contains toxic cardiac glycosides, which can cause heart attacks.[79,152] *Cerbera manghas* is not as toxic as some other species in the genus, however, which are notorious for their use in suicides and murders, and as a tribal 'ordeal poison' in which guilt is determined by invoking supernatural agents.[153] The generic name *Cerbera* is apparently derived from Cerberus, the three-headed dog of Greek mythology that guarded the gates of Hades, in reference to its reputation as a highly poisonous plant.[74]

The flowers have a tubular corolla (2.5–4 cm long) composed of five fused white petals, with a reddish ring encircling the apex of the tube.[151] As is common in many Apocynaceae species, the petal lobes (1.5–2.5 cm long) are characteristically twisted, overlapping on one side.[151] There are five stamens in each flower, attached to the top of the corolla tube.

The fruits (up to 6 cm long) are red and fleshy when ripe, features that are suggestive of animal frugivory and seed dispersal. Although there are several reports of *Cerbera manghas* fruits being eaten by bats,[154] the species is commonly dispersed by seawater,[155,156] with fruits often found washed up on shore after the outer fleshy part of the fruit (the pericarp) has rotted away, exposing the fibrous inner layer.[157]

Plate 28. *Cerbera manghas.* (A) Flowering branch, showing a terminal cluster of leaves and flowers. (B) Immature green fruit. (C) Mature red fruit (left), and fibrous remnant of a fruit after the fruit wall has rotted (right), commonly found washed ashore.

Cinnamomum camphora (L.) J. Presl

(= Laurus camphora L.)

—— ༄ ——

Cinnamomum camphora (Camphor tree; 樟; Lauraceae)[158] is an impressive species, with some individuals growing to 30 m in height and living up to 1,000 years.[80] The species is commonly found near villages and in *fung shui* woods,[1,30] leading botanists to recognise that it is not native to Hong Kong[10] although it has now become fully naturalised.[1]

The species is widely cultivated for its fragrant wood and as a source of camphor, a terpenoid compound that has a wide variety of culinary, medicinal and insecticidal uses. Camphor is readily absorbed through the skin, acting as an anaesthetic and reducing the body's thermal sensitivity:[159] camphor is accordingly an active ingredient in many commercially produced 'cooling' gels. Camphor also has antimicrobial properties and is toxic to many insects, leading to its use as a key component of mothballs. The wood is highly aromatic, and is ideal for the construction of chests and cabinets for the long-term storage of items that need to be protected from fungal growth and insect pests.

The flowers are small (*c.* 3 mm in diameter) and clustered into inflorescences that are 3.5–7 cm long.[158] As with most species in the family (*e.g., Litsea cubeba* and *Machilus velutina, q.v.*), the flowers have nine stamens arranged in three whorls of three, with an additional whorl of three sterile stamens ('staminodes'). Although the flowers are primarily pollinated by *Apis cerana* bees, they are also visited by many other insects, including calliphorid and syrphid flies, and papilionid and pierid butterflies.[61]

The fruits are small (6–8 mm in diameter) and fleshy, turning purplish-black at maturity.[158] Many different species of birds have been observed to eat the fruits, and these are likely to act as seed dispersers.[51,62] Seedlings of *Cinnamomum camphora* are comparatively rare, however, presumably due to the fact that almost all seeds are eaten by spotted doves (*Spilopelia chinensis*), which are seed predators rather than seed dispersers.[51] *Cinnamomum camphora* was successfully incorporated in the Government's reafforestation programme in the late nineteenth century,[25] and most of the older individuals encountered in Hong Kong today are therefore likely to have been planted rather than naturally seeded.

Plate 29. *Cinnamomum camphora.* (A) Flowering branch, showing clusters of small, white flowers. (B) Single flower. (C) Aggregates of small, purplish-black fruits.

Cratoxylum cochinchinense (Lour.) Blume

(= Cratoxylum ligustrinum (Spach) Blume; *Cratoxylum polyanthum* Korth.)*

——— ∽ ———

Cratoxylum cochinchinense (Yellow cow wood; 黃牛木; Clusiaceae or Guttiferae)[160] is a native tree species that is common in shrubland and early-stage secondary forests. The bark is smooth and peels away to reveal characteristic yellow, brown and green patches. The English and Chinese vernacular names are derived from the resemblance of the bark to the skin of local cattle.

Cratoxylum cochinchinense belongs to the same plant family as the cultivated mangosteen (*Garcinia mangostana*). The flowers in this family typically have five conspicuous petals and numerous stamens that are aggregated into ten clusters ('fascicles'), arranged in two whorls of five: the outer of these two whorls alternate with the petals and are sterile, forming structures known as 'fasciclodes' that swell at maturity, opening the flower by forcing the petals apart.[161] *Cratoxylum cochinchinense* shows some evolutionary modification of this pattern, however: the flowers are relatively small (*c.* 1 cm across) with five red, overlapping petals; the fertile stamens are aggregated into only three fascicles, presumably by fusion of adjacent fascicles (in a '2+2+1' arrangement);[162] and the sterile fasciclodes are similarly reduced to only three. The flowers are bisexual, with a compound pistil with three stigmas, derived from the fusion of three carpels. The flowers are pollinated by various bee species, but predominantly *Apis cerana*.[61]

The flowers of *Cratoxylum cochinchinense* have large sepals that persist after the flower has been fertilised, and are apparent even in mature fruits. The fruits are dry 'capsules' that split into three segments at maturity to release the seeds. Each seed possesses a lateral wing that assists in wind dispersal,[103,163] enabling the species to invade degraded areas that are rarely reached by animal-dispersed species. *Cratoxylum cochinchinense* is an important pioneer species as the trees provide perches for birds and hence encourage the deposition of seeds of other species that are carried by birds, thereby paving the way for subsequent ecological succession.[163]

Plate 30. *Cratoxylum cochinchinense.* (A) Flowering branch, showing clusters of small, red flowers. (B) Single flower. (C) Dissected flower, showing the following component organs (arranged left to right): fascicle of fertile stamens; sterile fasciclodes; compound pistil with three divergent stigmas; petal; and sepal. (D) Fruit capsules.

A

B

C

D

Cyclobalanopsis championii (Benth.) Oerst.
(= *Quercus championii* Benth.)

———— ᏰᎧ ————

Cyclobalanopsis championii (Champion's oak; 嶺南青岡; Fagaceae)[142] is a large tree species, growing to *c.* 20 m in height, that is locally common in relict woodlands in remote ravines in Hong Kong.[1] The genus *Cyclobalanopsis* is very closely related to the 'true' oaks, which are classified in the genus *Quercus*, with many contemporary taxonomists preferring to regard *Cyclobalanopsis* as a subgenus within *Quercus*, noting that it represents a distinct evolutionary lineage of Asian oaks.[164]

The wind-pollinated flowers are small, unisexual, and aggregated in separate inflorescences that are borne concurrently on the same tree. The male inflorescences (4–8 cm long) are pendent, whereas the female counterparts are shorter (*c.* 4 cm long) and held erect.[142] Each female flower is surrounded by an 'involucre' derived from fused, overlapping bracts.

Castanopsis and *Quercus* fruits are easily recognisable as acorns, in which a solitary nut is nested within a cup-like structure known as the 'cupule' that is likely to be evolutionarily derived from fused inflorescence branches.[143] As with other Fagaceae species in Hong Kong (*e.g.*, *Castanopsis fissa* and *Lithocarpus glaber*, *q.v.*), these fruits have evolved for dispersal by 'scatter-hoarding' rodents, which disperse and store the seeds for later consumption. Rodents capable of such dispersal are no longer present in Hong Kong,[139] however, and as a result the tree species that are dependent on this dispersal mechanism are rare in secondary woodlands.

The nut within the acorn has a tough shell that protects the embryonic seedling prior to seed germination. Evolutionary pressures have achieved a delicate balance in the strength of the nut: nuts that are too weak would merely result in widespread consumption by rodents, whereas nuts that are too tough would fail to attract rodents and hence would not be dispersed.

Plate 31. *Cyclobalanopsis championii.* (A) Flowering branch, showing pendent male inflorescences. (B) Leaves (upper and lower surfaces). (C, D) Male inflorescences. (E) Acorn, with a solitary nut nested within its cupule.

A

E

B

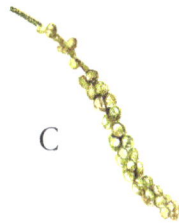

C

D

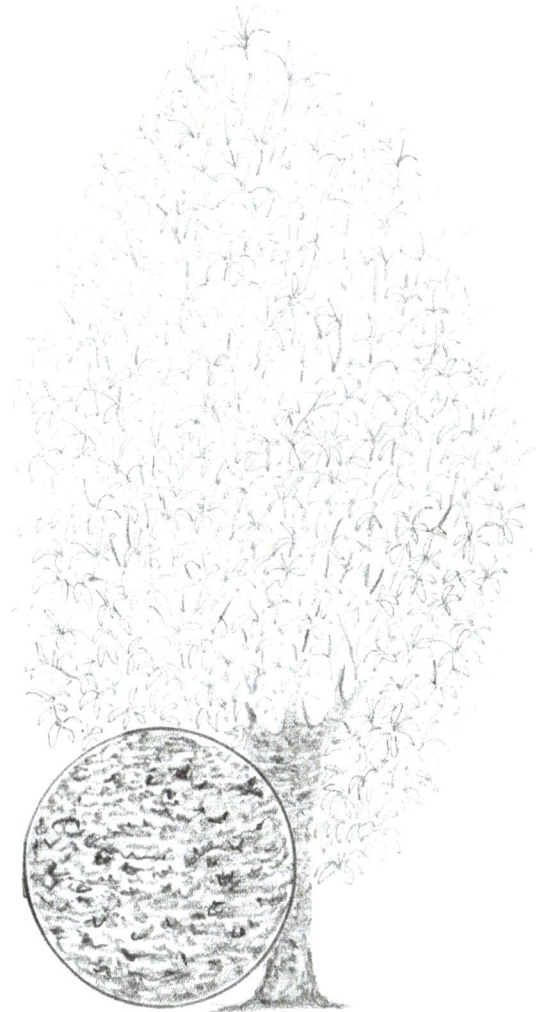

Dalbergia assamica Benth.

(= *Dalbergia balansae* Prain)

—— ✦ ——

Dalbergia assamica (South China rosewood; 南嶺黃檀; Fabaceae)[165] is native to southwest China, Indochina and northeast India. It is rare in the wild in Hong Kong, and although reported from Tai Mo Shan it may have been planted there: it has often been cultivated as an ornamental elsewhere in the territory because of its large inflorescences (up to 10 cm long) with attractive purplish-white flowers. Three of the six *Dalbergia* species in Hong Kong are climbers and one is a shrub: the two true tree species, *Dalbergia sissoo* (introduced from India) and *Dalbergia assamica*, are easily distinguished as the former has only 3–5 leaflets per leaf, whereas *Dalbergia assamica* has up to 27.

The flowers are typical of other 'papilionoid' (or 'pea-flowered') members of the family Fabaceae (*e.g.*, *Delonix regia* and *Ormosia emarginata*, *q.v.*) in being bilaterally symmetrical, with an enlarged upper petal (the 'standard'), paired lateral 'wings' and two lower 'keel' petals. Each flower has 10 stamens that are fused to form two bundles of five, and a single carpel. The wind-dispersed fruit pods are relatively small (5–6 × 2–2.5 cm), with 1–2 (rarely 4) seeds. The pods ripen in the late summer and often remain on the tree until the new flush of leaves develops the following spring.[157]

Many *Dalbergia* species—including the South American species *Dalbergia nigra* ('Brazilian rosewood')—are harvested for their valuable fragrant timber. In China, the most sought-after rosewood is 'huanghuali,' which is obtained from *Dalbergia odorifera* and was widely used for making high-quality furniture during the late Ming and Qing dynasties.[166] Many *Dalbergia* species have inevitably been over-exploited, with the Chinese species *Dalbergia odorifera* now regarded as vulnerable to extinction.[167]

Plate 32. *Dalbergia assamica.* (A) Flowering branch, showing inflorescences of purplish-white flowers. (B) Leaf. (C) Flowers. (D) Dry fruit pod with two seeds.

Daphniphyllum calycinum Benth.

———— ❧ ————

Daphniphyllum calycinum (Daphniphyllum; 牛耳楓; Daphniphyllaceae)[168] is a shrub or small tree species (growing to 5 m) that is common in Hong Kong shrublands and along the margins of secondary forests.[1] The leaves are thick and leathery and although alternately positioned are often crowded towards the apex of the branch and appear to be whorled. Individual trees bear flowers of only one sex (to ensure that individuals cannot self-fertilise, thereby promoting genetic diversity of the seeds produced). The flowers are very highly reduced, with both sexes lacking petals and with the sepals reduced to small lobes. The male flowers have nine or ten stamens, each with a very short filament and conspicuous oblong anthers, whilst the female flowers have two fused carpels.

The fruits are small (*c.* 1 cm long) single-seeded 'drupes' (resembling berries but with a tough protective inner fruit wall). They are blue with a waxy or powdery surface bloom and are very attractive to birds, which are the main frugivores[62] (although civets have also been recorded as seed dispersers[53]). Comparative studies of the longevity of fleshy fruits have revealed that although the ripe fruits of *Daphniphyllum calycinum* generally do not persist on the shrub for more than a week before being consumed by birds, the fruits persist for three months if the fruits are artificially enclosed in protective bags to prevent access by frugivores.[169] This longevity is presumably the result of effective anti-microbial defence compounds in the fruit flesh. *Daphniphyllum calycinum* is therefore unusual since fruits that display a significant resistance to microbial infection are generally less attractive to frugivores due to the presence of antimicrobial compounds; this suggests that species such as *Daphniphyllum calycinum* might be of potential pharmaceutical interest for future studies aimed at increasing the shelf-life of commercially important fruit crops.[169]

Although aluminium is common in soils, it is generally present as harmless aluminium oxides or aluminosilicates; soil acidification solubilises the aluminium, however, forming the trivalent cation (Al^{3+}) that is toxic to many plants, even in very small concentrations.[170] *Daphniphyllum* species have evolved mechanisms to chemically bind and accumulate aluminium in their leaves,[74] thereby allowing growth on aluminium-rich acidic soils.

Plate 33. *Daphniphyllum calycinum.* (A) Flowering branch, showing clusters of male flowers. (B) Male inflorescnce, viewed from above. (C) Male flower. (D) Fruiting branch, showing bluish drupes with waxy or powdery surface bloom. (E) Drupes with persistent sepals.

A

B   C

E

D

Delonix regia (Bojer) Raf.

——— ✧ ———

Delonix regia (Flame-of-the-forest; 鳳凰木 or 金鳳; Fabaceae)[119] is one of the most charismatic tropical tree species, with flamboyant clusters of large scarlet flowers. It was first discovered in eastern Madagascar in 1828 by Wenzel Bojer, who introduced it into cultivation in Mauritius;[171] this was the source for subsequent cultivation elsewhere, and the species is now widely planted as an ornamental tree throughout wet tropical regions. For many years *Delonix regia* was believed to be extinct in the wild, but was rediscovered in 1932.[172] Although often described as extremely rare in its natural habitat, several other naturally occurring populations have since been discovered.[173]

Despite its extensive cultivation, comparatively little is known of its floral biology: *Delonix regia* is a classic example of a 'well-known' species that is paradoxically only imperfectly understood.[171] Reports of floral visitors (birds and butterflies) are based on cultivated individuals outside its natural range; it is thought to be predominantly bird pollinated, however, and to have evolved from butterfly-pollinated ancestors with yellowish-white flowers.[174]

The flowers are bilaterally symmetrical, with the uppermost petal (the 'standard') distinctly striped. Each flower opens in the early morning and lasts for two days before the petals begin to abscise: on the first day, the standard is fully expanded, the flower produces copious nectar, and the anthers split open to release the pollen; and on the second day, the standard becomes folded, the nectar becomes less abundant, and the stigma becomes receptive.[171] The stamens therefore mature before the carpels (a phenomenon known as 'protandry'): this temporal separation of male and female function in the flower has evolved to minimise the potential for self-pollination within the flower and hence increase the chance of producing genetically diverse progeny. Individual flowers within each inflorescence furthermore develop sequentially and at least one day apart, hence avoiding the co-occurrence of multiple flowers at the same developmental stage.

Plate 34. *Delonix regia.* (A) Flowering branch, showing bilaterally symmetrical flowers with distinctly pigmented upper 'standard' petal. (B) Whorl of ten stamens, surrounding the carpel.

A

B

Dimocarpus longan Lour.

(= *Euphoria longan* (Lour.) Steud.; *Nephelium longana* Cambess.)

——— ❧ ———

Dimocarpus longan (Longan or Dragon's eye; 龍眼 or 桂圓; Sapindaceae)[175] is an important fruit crop in Southeast Asia,[176] and the species is accordingly widely planted throughout the region. In Hong Kong and southern China, *Dimocarpus longan* is commonly encountered in traditionally maintained *fung shui* woods close to villages, and is now regarded as semi-naturalised.[1] Longans belong to the same family as lychees (*Litchi chinensis*), and the two species can be hybridised despite belonging to different genera.[177]

Dimocarpus longan bears large inflorescences with numerous small flowers. The flowers are generally unisexual, with five brownish-yellow sepals and five white petals that are essentially the same length as the sepals. Male flowers have eight protruding stamens, whereas female flowers have a 2–3-lobed pistil. The maturation of flowers in the inflorescence follows a specific sequence: the first to mature possess stamens but lack a pistil (*i.e.*, are functionally male); these are followed by flowers with sterile stamens ('staminodes') and a fertile pistil (*i.e.*, functionally female); which are in turn followed by flowers that have fertile stamens and a rudimentary pistil (*i.e.*, functionally male).[177] This temporal transition of sexual expression between flowers is just one of many adaptations shown by plants to reduce the possibility of self-pollination and hence self-fertilisation, thereby promoting genetic mixing and avoiding the negative consequences of inbreeding.

In addition to its uses as an edible fruit crop,[176] *Dimocarpus longan* is used as a source of timber[80] and as a traditional medicine for reducing pain and swelling.[178]

Plate 35. *Dimocarpus longan.* (A) Flowering branch, showing large inflorescences comprising numerous small flowers. (B) Solitary male flower, with protruding stamens. (C) Cluster of fruits. (D) Open fruit, opened to show seed. (E) 'Eye-like' seed embedded within the fleshy inner fruit wall

Diospyros morrisiana Hance

———— ✑ ————

Diospyros morrisiana (Morris's persimmon; 羅浮柿; Ebenaceae)[179] is a deciduous tree with a distinctive blackish bark, common in shrubland and young secondary forests in Hong Kong.[1] The lower surface of the leaves often have conspicuous extra-floral nectaries;[180] these nectaries produce a sugary exudate that presumably encourages ant populations, which are likely to act as a deterrent against potential herbivores.

Diospyros morrisiana flowers are unisexual, with male flowers aggregated in clusters of two or three, whereas female flowers are solitary. Each flower has four creamy-white petals that are fused into an urn-shaped corolla tube with reflexed apical lobes. The male flowers have 16–20 hairy stamens attached to the base of the corolla tube, whereas the female flowers have a fused pistil and six sterile stamens ('staminodes'). The flowers are visited by bees (especially *Apis cerana*) and wasps.[61]

The fruits of *Diospyros morrisiana* are fleshy berries that turn yellow at maturity. They are too large (*c.* 19 mm in diameter) to be swallowed whole by most birds: although birds sometimes peck at the flesh of the fruit, they avoid the large seeds (10–14 × 5–7 mm) and so are not responsible for seed dispersal.[2] In Hong Kong, the fruits are eaten by civets, including the Small Indian civet (*Viverricula indica*),[62,110,181] although elsewhere in Asia macaques[182,183] and flying foxes[184] are known to disperse the seeds.

Although *Diospyros morrisiana* is of little commercial importance, it is closely related to *Diospyros kaki*, which is cultivated in China and Japan for its fruits (Kaki, Chinese or Japanese date plum, or Persimmon),[185] and several *Diospyros* species that are cultivated for their timber (ebony).[186]

Plate 36. *Diospyros morrisiana.* (A) Flowering branch. (B) Fruiting branches, with slightly immature fruits that become yellow as they ripen. (C) Single fruit, sectioned longitudinally.

Elaeocarpus chinensis (Gardner & Champ.) Hook. f. & Benth.

———— ❧ ————

Elaeocarpus chinensis (Chinese Elaeocarpus; 中華杜英 or 野杜英; Elaeocarpaceae)[187] is a small native tree species (growing to *c.* 7 m in height) that is locally common in shrubland and early successional secondary forests.[1] The family Elaeocarpaceae originated in the southern hemisphere, with most genera (including the two Hong Kong representatives, *Elaeocarpus* and *Sloanea*) showing greatest diversity in New Guinea;[188] this contrasts with most elements of our local flora, which show strong biogeographical links with the northern hemisphere.

Elaeocarpus chinensis flowers are borne in small pendent greenish-white clusters, with morphologically similar sepals and petals (slightly elongated, *c.* 3 mm long). The flowers are generally bisexual, with 8–10 stamens and a pistil derived from the fusion of two carpels, although male flowers (lacking a pistil) are also reported.[187] Each flower has a glandular disc that produces nectar that functions as a reward to pollinators: local studies have demonstrated that the flowers are primarily visited by *Apis cerana* bees.[61]

The fruits are small ellipsoid 'drupes,' with a bluish fleshy sugar-rich outer layer and a tough inner fruit wall surrounding the solitary seed. The fruits are therefore ideally adapted for dispersal by birds:[62] their size (*c.* 7 mm wide) is within the beak gape range of local frugivorous birds; their colour is within the range perceived by birds; and the seeds are small enough (*c.* 4.5 mm in diameter) to be swallowed by the birds and yet are protected by the tough inner fruit wall as they pass through the bird's digestive tract. The fruits of other species of *Elaeocarpus* are reported to be eaten by a great diversity of animals,[68] and it is therefore not surprising that *Elaeocarpus chinensis* fruits are also reported to be dispersed by civets.[181]

Plate 37. *Elaeocarpus chinensis.* (A) Flowering branch, showing clusters of small, pendent flowers. (B) Single flower. (C) Fruiting branch, showing bluish drupes.

A

B

C

Endospermum chinense Benth.

——— ৎৎ ———

Endospermum chinense (Endospermum; 黃桐; Euphorbiaceae)[77] is a large (up to *c.* 35 m), fast-growing and light-demanding pioneer tree species that is locally common[1] in Hong Kong. The leaves are typically crowded towards the apex of the young branches and are relatively large (8–20 × 4–14 cm) and borne on a long petiole (4–9 cm).[77] The lower surface of the leaf is densely covered with short hairs. As with many other species in the Euphorbiaceae family (including *Aleurites moluccana*, *Macaranga tanarius*, *Mallotus paniculatus*, *Sapium discolor* and *Vernicia montana*, *q.v.*), the petiole bears a conspicuous pair of glands at the base of the lamina. These extrafloral nectaries are likely to encourage colonies of ants that deter potential herbivores.[189] Other species of *Endospermum* (especially *Endospermum moluccanum* and *Endospermum myrmecophilum*) have evolved a considerably more complex association with ants in which the ants create a nest site by excavating cavities in branches by removing the central pith tissue.[189,190]

Individual trees are unisexual, bearing either male (staminate) or female (pistillate) flowers. This separation of sexes (known as 'dioecy') prevents self-pollination and hence self-fertilisation, and therefore promotes the mixing of genes in the seeds that are produced. The flowers are borne in branched inflorescences growing from the axils of leaves. As is typical for the Euphorbiaceae, *Endospermum chinense* flowers are highly reduced, with a small cup-shaped fused calyx but lacking any petals. The male flowers have 5–12 stamens borne on an elongated receptacle, whereas the female flowers have a 2–3-chambered pistil.[77]

Endospermum chinense fruits are fleshy 'drupes' (berry-like fruits with a tough inner fruit wall surrounding the seeds), capped with a persistent stigma.

Plate 38. *Endospermum chinense.* (A) Flowering branch, with branched inflorescences. (B) Stem showing scars of fallen leaves. (C) Single leaf (lower surface), showing paired glands at the apex of the petiole. (D) Fleshy drupes with their persistent stigma.

Engelhardia roxburghiana Lindl. ex Wall.

(= *Engelhardia chrysolepis* Hance; *Engelhardia wallichiana* auct. non Lindl.)

─── ∾ ───

Engelhardia roxburghiana (Yellow basket willow or Roxburgh's Engelhardia; 黃杞; Juglandaceae)[191] are tall trees growing up to 30 m in height, with large compound leaves (12–25 cm long) comprising 3–5 pairs of leaflets. The flowers are unisexual and borne in elongated inflorescences, with both sexes borne on the same tree: individual flowers are very small, with a persistent three-lobed bract and a four-lobed perianth. The male flowers have 5–13 stamens, and the female flowers have a pistil that is derived from the fusion of two carpels, with an 'inferior' ovary.

The flowers are wind pollinated, and subsequently develop into globose nutlets (*c.* 4 mm in diameter). The nutlet is borne on a greatly enlarged three-lobed bract (the central lobe of which is 3–5 cm long) that assists in wind dispersal.[68]

The Juglandaceae family includes many commercially important nut trees, including walnuts (*Juglans* species) and hickory and pecan nuts (*Carya* species). It appears that two different seed dispersal mechanisms operate in the family—wind and animal dispersal. Phylogenetic reconstructions based on DNA sequence data have helped to elucidate important evolutionary transitions:[33,192] although it appears that the family was ancestrally wind dispersed, there were at least four independent origins of the structural wings essential for such dispersal, derived from the trilobed bract, bracteoles, sepals or the fusion of these organs. Similarly, animal-assisted dispersal has been shown to have been derived on at least three occasions (including independent origins in *Juglans* and *Carya*). This is an excellent example of the way in which DNA-based molecular phylogenies can be used as a conceptual framework for integrating new interpretations of plant structure with novel ecological hypotheses.

Plate 39. *Engelhardia roxburghiana.* (A) Flowering branch, with elongated inflorescences. (B) Infructescence. (C) Fruit nutlet, borne on its three-lobed bract that assists with wind dispersal. (D) Senescent leaves.

A

B

C

D

Enkianthus quinqueflorus Lour.

———— ❧ ————

Enkianthus quinqueflorus (Enkianthus, Chinese New Year flower or Hanging bell flower; 吊鐘; Ericaceae)[193] is native to southern China and is relatively common in Hong Kong. The shrubs or trees are small (generally up to 3 m in height, although sometimes reaching 10 m), with conspicuous bell-shaped flowers that grow in pendent clusters. The corolla consists of five fused glossy pink petals, with each petal apex pale pink and reflexed. Each flower has 10 stamens with paired awn-like appendages.[194] The flowers are predominantly pollinated by birds (including the Japanese white-eye, *Zosterops japonica*), but are also visited by bees.[61]

Flowering occurs in winter, generally after the previous year's leaves have fallen and before the new years' foliage develops.[193] The beauty of the flowers and the fact that flowering often coincides with the Chinese lunar New Year has previously resulted in many trees being collected for ornamental use and sold in flower markets. The species has been legally protected in Hong Kong since 1913,[2] however, and there are now several well-established local populations.

Fruit development in *Enkianthus quinqueflorus* is very prolonged, with flowering occurring between January and March followed by fruiting between September and December.[193,195] The fruits are dry dehiscent capsules that split open along the midrib of each of the five carpels. Although the flowers are pendent, the resultant fruits are held erect on the tree; this may be advantageous for promoting wind dispersal of seeds.

Plate 40. *Enkianthus quinqueflorus.* (A) Flowering branch, showing the distinctive cluster of pendent bell-shaped flowers. (B) Corolla of opened flower, showing stamens (with paired appendages). (C) Fruit capsule and vegetative buds. (D) Open fruit capsule.

A

B

C

D

Eriobotrya japonica (Thunb.) Lindl.

(= Mespilus japonica Thunb.)

——— ❧ ———

Eriobotrya japonica (Loquat; 枇杷; Rosaceae)[196] is locally common in secondary forests in Hong Kong.[1] The species is native to south-eastern China, where it has been cultivated since antiquity for its edible fleshy, orange-yellow fruits (2–5 cm in diameter). The fruits have a sweet-sour taste and are either eaten fresh or preserved as a jam, with the seeds sometimes used as an almond-like flavouring in drinks and cakes.[197] Although loquats were reputedly introduced into cultivation in Japan in the twelfth century, it was not until the eighteenth century that European explorers introduced the species into cultivation across the Mediterranean; the species is now widely grown in many temperate and subtropical areas, especially in latitudes up to 35°.[198] China and Japan remain the main centres of loquat cultivation, although most of the crop is consumed within these countries rather than being traded internationally.[197]

Eriobotrya japonica bears multi-flowered inflorescences up to 20 cm in length with a densely hairy central axis and floral stalks. Individual flowers are fragrant and bee-pollinated,[199] with a fused calyx of five sepals and five white petals (5–9 × 4–6 mm), 20 stamens and a five-chambered pistil.[196] As with many other members of the Rosaceae family (including *Photinia benthamiana, q.v.*), *Eriobotrya japonica* flowers have an 'inferior' ovary in which the ovary is surrounded and protected by an extension of the floral receptacle known as the 'hypanthium'. The edible fruit flesh in loquats is derived from this hypanthium, and the fruit (known as a 'pome') resembles a miniature apple or pear, which belong to the same family.

Eriobotrya japonica fruits are known to be eaten by bats and birds,[68] which are effective seed dispersers (although many of these reports are from outside of the native range of the species). The seeds have been inferred to be dispersed in Hong Kong by civets.[62]

Plate 41. *Eriobotrya japonica.* (A) Flowering branch, with inflorescence of small, white flowers and densely hairy floral stalks. (B) Lower surface of leaf. (C, D) Flowers. (E) Fruiting branch with yellowish 'pomes'. (F) Seeds.

Exbucklandia tonkinensis (Lec.) H. T. Chang

(= *Bucklandia tonkinensis* Lec.; *Symingtonia tonkinensis* (Lec.) Steenis)

——— ℰ℀ ———

Exbucklandia tonkinensis (Exbucklandia; 大果馬蹄荷; Hamamelidaceae)[200] is widely distributed in southern China and Indochina, but is very rare in Hong Kong, where it is only known from Sunset Peak on Lantau.[98] The Agriculture, Fisheries & Conservation Department of the Hong Kong Government has accordingly initiated propagation and *ex situ* conservation of the species locally.[98]

Exbucklandia tonkinensis grows to *c.* 30 m with a massive trunk that yields a high-quality timber, and bears ovate or three-lobed leaves that are smooth and lack hairs. The flowers are very small and tightly aggregated in clusters of 7–9 in the leaf axils. Since sepals and petals are absent, individual flowers are reduced to their reproductive organs, with 10–15 stamens and a pistil derived from two fused carpels. Unlike most flowering plants—which have stamens with four distinct pollen-bearing chambers ('tetrasporangiate' stamens)—the stamens of *Exbucklandia* species only have two chambers ('bisporangiate').[201] Although little is known of the pollination ecology of *Exbucklandia tonkinensis*, the lack of petals suggests that it is wind pollinated: studies of other members of the Hamamelidaceae indicate several parallel shifts from insect pollination to wind pollination associated with petal loss.[202]

After fertilisation, the carpels develop into small dry fruiting capsules (10–15 × 8–10 mm) that split open apically to release the seeds. Members of the Hamamelidaceae family have two distinct seed dispersal mechanisms: some species (including species of *Disanthus* and *Mytilaria*, which are closely related to *Exbucklandia*) have a ballistic mechanism in which seeds are ejected several metres from the fruit; whilst other species (including *Exbucklandia tonkinensis*) develop small, winged seeds that are wind dispersed.[203]

Plate 42. *Exbucklandia tonkinensis.* Flowering and fruiting branches, with closely appressed flowers and fruit capsules.

Ficus variolosa Lindl. ex Benth.

—— ❧ ——

Ficus variolosa (Mountain fig or Varied-leaf fig; 變葉榕; Moraceae)[109] belongs to an extremely diverse genus characterised by tiny, highly reduced flowers that are aggregated in a complex structure known as a 'syconium.' These syconia superficially resemble fruits, but are actually completely invaginated, hollow inflorescences in which the tiny flowers are borne on the inner surface of the enclosed chamber. Each syconium has a small distal aperture (the 'ostiole') that enables entry of the pollinators.

Pollination of fig species is particularly fascinating as it involves a complex 'mutualistic' symbiosis in which a unique species of fig wasp (belonging to the family Agaonidae) is associated with each *Ficus* species, and in which the two species are mutually inter-dependent and cannot survive without the other.[92,204] Specific details of the pollination mechanism differ between *Ficus* species, although it is invariably the female fig wasp that achieves pollen transfer between different fig trees.[35] In *Ficus variolosa*, some syconia are functionally female, containing only female flowers that have fertile ovaries with viable eggs; other syconia are functionally male and contain a combination of male flowers (located close to the ostiole) and sterile flowers with non-functional ovaries (deeper inside the inflorescence chamber).[109] The female wasps enter the syconium via the ostiole. In female syconia, the wasps attempt to lay their eggs in the floral ovaries, but are unsuccessful since the styles on the carpels are too long and prevent the wasps from reaching the ovaries; pollen is nevertheless likely to be transferred from the wasps to the stigmas during this process, resulting in successful pollination and seed set. A different scenario unfolds in male syconia, however, as the sterile flowers have shorter styles that enable the wasps to successfully lay their eggs in the non-functional ovaries; these ovaries develop to form gall-like structures that enclose the developing wasp larvae. The fig wasps subsequently emerge from the galls in two cohorts. The first cohort of wasps to emerge are wingless males that rapidly complete their brief life cycle without ever leaving the syconium, fertilising female fig wasps that are still in their galls, and then digging an escape tunnel for the females through the wall of the syconium before they die. The mated female wasps subsequently emerge from their galls and escape via the tunnels created by the male wasps; by this stage, the male flowers in the syconium have matured and released their pollen grains, ensuring that the female fig wasps are able to transfer pollen to other syconia.

Plate 43. *Ficus variolosa.* Flowering branch, showing fruit-like syconia.

Garcinia oblongifolia Champ. ex Benth.

———— ❧ ————

Garcinia oblongifolia (Lingnan Garcinia; 嶺南山竹子 or 黃牙果; Clusiaceae or Guttiferae)[160] is a very common, native tree species, growing to *c.* 15 m in height.[1] It belongs to the same plant family as *Cratoxylum cochinchinense* (*q.v.*), with which it shares several similarities in floral structure, especially with regard to the fusion of stamens. *Garcinia oblongifolia* flowers are unisexual, with both sexes of flower borne concurrently on the same tree. Each flower has four small, rounded sepals (3–4 mm in diameter), and four yellowish-white, elongated petals (7–9 mm long).[160] The male flowers have many stamens, aggregated into a four-sided fleshy bundle and lacking any vestigial female organs; the female flowers, in contrast, have a fused pistil and bundles of 'staminodes' (sterile stamens).

Despite the similarities in floral structure between *Garcinia oblongifolia* and *Cratoxylum cochinchinense*, the fruits of the two species are strikingly different, with fleshy berries in the former and dry capsules in the latter. *Garcinia oblongifolia* berries (2–4 × 2–3.5 cm) are structurally similar to the cultivated mangosteen (*Garcinia mangostana*), which belongs to the same genus. Although the berries are edible, the Chinese vernacular name 黃牙果 (which translates as 'fruit causing yellow teeth') implies that some caution is required!

The berries have a tough inedible skin that has to be removed before the sucrose-rich[63] fleshy pulp can be accessed. *Garcinia oblongifolia* fruits are eaten by macaques, which initially remove part of the fruit skin using their incisors, and then scoop out the flesh and seeds with their teeth.[2,110] The seeds are too large to be swallowed and hence are spat out rather than ingested; they are nevertheless carried over a considerable distance since the fruit mass is temporarily stored in cheek pouches.[205] Whilst macaques are undoubtedly effective seed dispersers in Hong Kong,[62] the extensive distribution of *Garcinia oblongifolia* in areas where macaques are not resident suggests that alternative dispersal agents must exist.[110]

Plate 44. *Garcinia oblongifolia*. (A) Flowering branch. (B) Fruiting branch. (C) Dissected fruits, showing the thick, tough peel that has to be removed by frugivores before they can access the fleshy pulp.

A

B

C

Glochidion zeylanicum (Gaertn.) A. Juss.

(= Glochidion hongkongense Müll. Arg.; *Glochidion littorale* auct. non Blume)

—— ℰℐ ——

Glochidion zeylanicum (Hong Kong abacus plant or Sri Lankan Glochidion; 香港算盤子; Phyllanthaceae)[77] is a widespread tree species in Southeast Asia and is common in lowland forests in Hong Kong and southern China. It has highly reduced unisexual flowers, with both sexes of flower borne on the same tree simultaneously. Each flower has six sepals but lacks petals; the male flowers have 5–6 stamens that are fused to form a column, whereas the female flowers have a united pistil with styles that are similarly fused into a column.

The flowers are sexually functional at night and are pollinated nocturnally by female moths in the genus *Epicephala* (family Gracillariidae).[206,207] Considerable specificity is apparent in the moth-tree association, with each *Glochidion* species pollinated by a single moth species; this specificity has been shown to be mediated by the highly specific floral scents that are emitted to attract the moths.[208] This is a rare example of a pollination mutualism, similar to that described for fig trees (*e.g., Ficus variolosa, q.v.*). The female *Epicephala* moths lay their eggs in the floral ovaries, and during this process transfer pollen that was inadvertently collected from previous visits to other flowers. The eggs laid in the floral ovaries subsequently hatch and the larvae consume the developing seeds; although these ovaries therefore cannot produce seeds, other ovaries that were not used as egg-laying sites by the moths are able to set seed. The relationship between the *Epicephala* moth and *Glochidion* tree is therefore clearly mutualistic: the moth benefits from a guaranteed food source for its developing larvae; and the tree benefits from the highly assured level of pollination success. The success of this mutualism comes at a cost, however, since neither the tree nor the moth can successfully reproduce in the absence of the other.

The fruits are red and have deep ridges running from the apex to the base, along which the fruit splits open at maturity to release the seeds. The English vernacular name 'abacus plant' is presumably derived from the resemblance of these capsules to beads on an abacus.

Plate 45. *Glochidion zeylanicum.* (A) Flowering branch, showing highly reduced flowers. (B) Male flower. (C) Female flower. (D) Fruiting branch, showing red fruit capsules. (E) Fruit capsule.

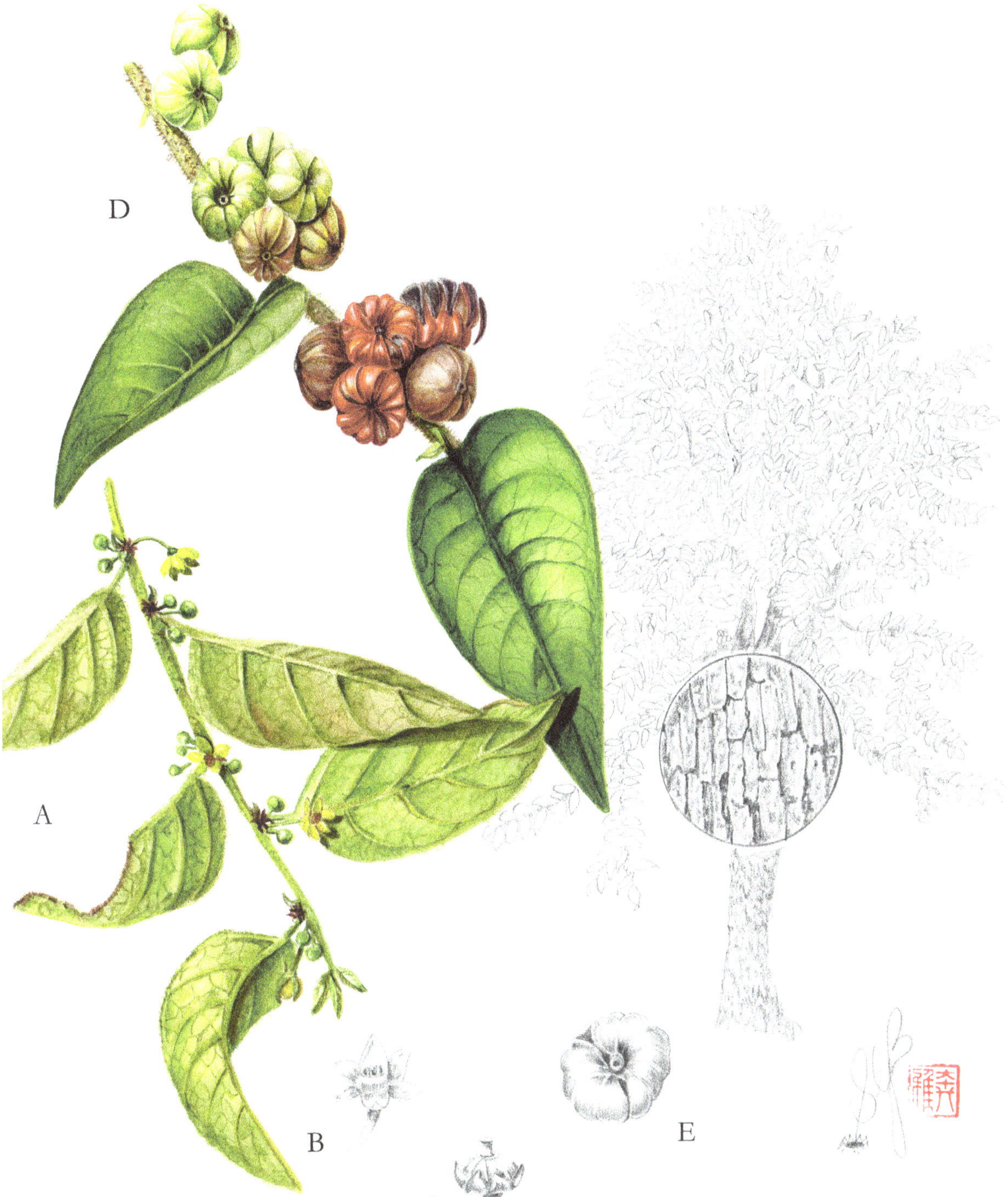
D
A
B
E
C

Gmelina chinensis Benth.

———— ∾ ————

Gmelina chinensis (Gmelina; 石梓 or 華石梓; Lamiaceae)[132] has a relatively localised distribution in Hong Kong, although it is locally common in Lantau; the species is nevertheless regarded as being vulnerable to extinction.[98] The genus was traditionally classified in the family Verbenaceae—the approach adopted, for example, in the *Flora of China*[209] and *Flora of Hong Kong*.[132] Molecular phylogenetic analysis of DNA sequence data, however, has revealed that the Verbenaceae does not represent a single evolutionary lineage and has therefore had to be recircumscibed;[210] this reclassification has necessitated the transfer of several genera, including *Gmelina*, to the closely related family Lamiaceae (also known as the Labiatae).

The flowers have five petals that are fused basally into a tube with conspicuous petal lobes; the outside of the corolla is yellow, and the inside is pink-purple. The flowers are bilaterally symmetrical, with a three-lobed lower petal.[211] There are four stamens per flower, arranged in two pairs of slightly different lengths. The style is long (25–27 mm), and the unequally two-lobed stigma is held above the anthers of the stamens. The flowers are likely to be bee-pollinated since this mechanism has been observed in other species in the genus:[212] the bees presumably brush past the anthers and collect pollen as they probe into the flower and subsequently deposit the pollen on the stigma from another flower visited.

Various organs on the tree bear extra-floral nectaries, including the base of the leaves, the calyx and the fruit.[49,211] These glands presumably encourage ants that may act as a deterrent against potential herbivores.

Plate 46. *Gmelina chinensis.* (A) Flowering branch, showing clusters of conspicuous flowers. (B) Corolla tube, opened to show reproductive organs. (C) Calyx of small, fused sepals. (D) Stamens.

Heritiera littoralis Aiton

—— ☙ ——

Heritiera littoralis (Coastal Heritiera; 銀葉樹; Malvaceae)[213] is a medium-sized tree that is restricted to coastal regions, including mangroves; although it is widespread across large areas of the Indian and Pacific Oceans, it is rare in Hong Kong, and only recorded from Tai Po Kau, Kat O Chau, Lai Chi Wo and Yung Shue O.[1,213] The base of the trunk in older trees has well-developed buttresses that provide additional support on soils that are unstable because of periodic tidal inundations.

Heritiera littoralis flowers are borne in loose inflorescences and lack petals, although they possess a brownish-red cup-shaped calyx composed of 4–6 fused sepals. Each flower is unisexual with both sexes of flower borne on the same tree, and insect pollinated.

The fruits that develop after fertilisation are large (*c.* 6 × 3.5 cm), with a prominent raised 'keel' that runs the length of the fruit. The combination of the cork-like fruit wall and the presence of an air-gap surrounding the seed ensure that the fruits remain buoyant in water,[68] hence providing a convincing explanation for the widespread geographical range of the species. The English botanist and former Director of the Singapore Botanic Garden, Henry N. Ridley (1855–1956), was a pioneer of plant biogeography and mechanisms of plant dispersal. In his seminal book *The Dispersal of Plants Throughout the World* (1930),[68] Ridley suggested that the genus *Heritiera* provides an excellent demonstration of evolutionary changes in seed dispersal mechanism: from wind-dispersed species in *Tarrietia* (now treated as species of *Heritiera*[214]) with large broad wings that enable rotation of the fruit whilst it is falling from the tree; to *Heritiera* species with a reduced wing that are dispersed by rivers; to *Heritiera littoralis*, which has an even narrower wing but which, as noted above, has other adaptations enabling marine dispersal.

Plate 47. *Heritiera littoralis.* (A) Flowering branch. (B, C) Inflorescence and single flower, showing fused red sepals. (D) Fruiting branch, with immature fruits. (E) Mature fruit, showing prominent raised keel. (F) Large buttresses evident at the base of the trunk of older trees.

Hibiscus tiliaceus L.

———— ❦ ————

Hibiscus tiliaceus (Cuban bast, Sea coast mallow or Sea Hibiscus; 黃槿; Malvaceae)[215] is salt tolerant[216] and grows in coastal thickets behind beaches, although it sometimes also occurs further inland. The flowers are large and showy, with petals that are yellow with a dark red base, and are primarily visited by bees. As with most members of the mallow family, the filaments of the stamens are fused into a cylinder that surrounds the united style of the fused carpels.

The flower develops into a dry dehiscent fruit capsule at maturity. Each fruit capsule produces 50–70 small seeds (3×5 mm) with internal air chambers that ensure buoyancy:[217,218] experimental studies have demonstrated that these seeds can float in seawater and remain viable for more than three months.[155] This evolutionary adaptation has been key to the success of the species in colonising new geographical areas, including isolated oceanic islands: *Hibiscus tiliaceus* is very widespread in the tropics and subtropics, occurring in coastal regions in Africa (including both Atlantic and Indian Ocean coasts), Asia, Australasia and the western Pacific.[217]

Studies of the evolution of *Hibiscus tiliaceus* based on the analysis of DNA sequence data[217] have revealed that it is likely to have undergone multiple speciation events, giving rise to many allied species, including *Hibiscus pernambucensis*, which is widespread in both the Pacific and Atlantic littoral zones of North, Central and South America, where *Hibiscus tiliaceus* does not grow. Although both *Hibiscus tiliaceus* and *Hibiscus pernambucensis* are capable of extensive long-distance dispersal by sea-drift of seeds, it is likely that oceanic currents in the Atlantic and eastern Pacific have created significant barriers to dispersal and hence to genetic exchange.

Plate 48. *Hibiscus tiliaceus.* (A) Flowering branch, showing the large, showy yellow flowers with deep red centres. (B) Fused stamen tube surrounding united style, with five stigmas visible at the top. (C, D) Dehisced fruit capsules, with seeds.

Homalium cochinchinense (Lour.) Druce

(= *Homalium fagifolium* Benth.)

——— ∾ ———

Homalium cochinchinense (Homalium; 天料木; Salicaceae)[219] is a common species in local shrubland and early-stage secondary forests. Although it can grow as a tree to 10 m in height, its growth-form is generally shrubby.[1]

The small, white flowers are borne in elongated pendent inflorescences. Each flower has 7–8 narrow sepals and an equal number of slightly wider petals. Each petal has a single stamen basally fused to it, and alternates with conspicuous orange-yellow glands. The pistil typically has three stigmas. Two distinct peak flowering periods are evident—early and late in the wet season—although the latter flowering period typically produces few or no fruits.[195] The flowers are pollinated by bees (especially *Apis cerana*)[61] that are presumably rewarded by nectar produced by the floral glands.

The genus *Homalium* has traditionally been classified in the family Flacourtiaceae, an approach that has been accepted in both the *Flora of China*[220] and *Flora of Hong Kong*.[219] This circumscription of the family is very heterogeneous, however, and phylogenetic analysis of DNA sequence data has suggested that it comprises two distinct evolutionary lineages.[221] The lineage that includes *Homalium* also includes members of the willow and poplar family, Salicaceae, which are characterised by 'catkins' of highly reduced, unisexual and generally wind-pollinated flowers. It therefore seems probable that an evolutionary reduction in floral morphology has occurred, with loss of sepals and petals, and that *Homalium* and closely related genera (including *Scolopia chinensis, q.v.*) have retained many of the ancestral character states, including conspicuous sepals and petals. Contemporary classification schemes normally only include taxonomic groups that have been shown to represent distinct evolutionary lineages: groups are therefore rejected if they represent more than one evolutionary lineage, or if they have another group nested within them. On this basis, *Homalium* and related genera have been transferred to the Salicaceae in most recent classifications.[12,13]

Plate 49. *Homalium cochinchinense.* (A) Flowering branch, showing long pendent inflorescences with numerous flowers. (B) Isolated flower. (C) Dissected flower, showing orange-yellow glands. (D) Pistil with three stigmas. (E) Stamen. (F) Fruits, with persistent calyx. (G) Autumn leaf.

Ilex rotunda Thunb.

—— ☙ ——

Ilex rotunda (Chinese holly or Panaceae holly; 鐵冬青; Aquifoliaceae)[222] is typical of most holly species in possessing clusters of small unisexual flowers, with the two sexes of flower borne on separate individuals—a phenomenon known as 'dioecy', which promotes genetic mixing in progeny by preventing self-fertilisation. Each flower has 4–6 petals that are basally fused. Male flowers have an equal number of stamens that are fused to the petal tube, and a rudimentary and non-functional ovary. Female flowers are similar, but with a functional ovary and sterile stamens that lack pollen. Most *Ilex* species show a male-biased sex ratio in populations, although the reasons for this remain obscure; the data for *Ilex rotunda* is rather equivocal, however, due to the inadequate population sizes sampled.[223]

As with other species in the genus, *Ilex rotunda* flowers are primarily pollinated by bees, which are attracted by nectar: *Apis cerana* was the most frequent floral visitor to populations in Hong Kong, with an average of 11–12 visits per hour.[223]

The fruits are bright red and fleshy, and are consumed by birds (especially the Light-vented bulbul, *Pycnonotus sinensis*, and the Japanese white-eye, *Zosterops japonica*), which are effective seed dispersers.[223] Although the main fruiting period for *Ilex rotunda* in Hong Kong is from September until January of the following year, some fruits remain on the tree for several further months, sometimes until April.[223] This persistence of fruits is likely to be due to the relative 'unattractiveness' of the fruits to frugivores due to their low sugar content (*c.* 24% of pulp dry mass, compared with *c.* 53% for other non-*Ilex* species with fleshy fruits),[223] and the presence of chemical defense against microbial infection.[169] *Ilex* seeds have rudimentary embryos when the fruits reach maturity,[224] and as a result the seeds can remain dormant in the soil seed bank for several years.[225]

Plate 50. *Ilex rotunda.* (A) Flowering branch, showing inflorescences of small white flowers. (B) Male and female flowers. (C) Fruits.

B

A

C

Illicium angustisepalum A. C. Sm.
(= *Illicium spathulatum* auct. non Y. C. Wu)

——— ❧ ———

Illicium angustisepalum (Lantau star-anise; 大嶼八角; Illiciaceae or Schisand-raceae)[226] is very rare and is only known locally from Lantau Peak (where it was first collected in 1905) and Sunset Peak, Lantau. It is often regarded as a Hong Kong endemic[98, 226] although this interpretation is not accurate if a broader circumscription of the species is accepted, inclusive of populations in Anhui and Fujian provinces in China.[227] The rarity of the species has nevertheless led to it being legally protected in Hong Kong.[98]

Illicium angustisepalum belongs to one of the earliest surviving evolutionary lineages of flowering plants and possesses many morphological characteristics that botanists interpret as ancestral.[33] The flowers are bisexual with numerous spirally arranged organs. The perianth consists of 22–24 weakly differentiated 'tepals,' with a gradual transition between outermost tepals that are sepal-like and innermost tepals that are petal-like. The stamens are flattened (often described as 'tongue-like'), with the pollen-bearing structures borne on the inner surface. The flowers have 11–13 unfused carpels that appear to be arranged in a whorl (although they are actually spirally arranged[228]). This contrasts with species in more derived evolutionary lineages, which generally have carpels that are fused together: fusion of carpels is advantageous as it enables the pollen tubes that carry the sperm to reach and fertilise any of the ovules in the flower rather being restricted to fertilising ovules within each unfused carpel.

After fertilisation, the carpels develop into a characteristic ring of discrete dry fruits, each of which splits open across the upper surface to expose a single glossy seed. It has been suggested that this fruit type—known as a 'follicle'—might be the ancestral condition in flowering plants: follicles have been described from *Archaefructus*,[229,230] which is one of the earliest flowering plant fossils (dating from the Early Cretaceous). Other researchers have disputed the phylogenetic position of *Archaefructus* and this interpretation of its fruit structure, however,[231,232] implying that the follicles in *Illicium* and its relatives may be derived.

Plate 51. *Illicium angustisepalum.* (A) Flowering branch, showing conspicuous white tepals. (B) Bark. (C) Immature fruit, shortly after fertilisation of the ovules. (D) Mature star-shaped fruit, composed of separate follicles. (E) Seeds.

Itea chinensis Hook. & Arn.

———— ☙ ————

Itea chinensis (Itea; 老鼠刺 or 鼠刺; Iteaceae)[233] is very common in Hong Kong shrublands and young secondary forests,[1] although as an early-successional species it does not persist into older closed-canopy forests.[103] Individuals grow to *c.* 15 m in height and bear terminal or axillary inflorescences (up to 8 cm in length), composed of numerous small flowers. The flowers have five sepals that are fused into a cup that surrounds an equal number of free, white petals. Each flower is bisexual, with five stamens that alternate with the petals, and a two-chambered pistil. After fertilisation, the pistil develops into a dry capsule that separates into two segments, each adorned with a persistent style and its rounded stigma. The splitting of the capsule at maturity allows the seeds, which are wind-dispersed,[103] to escape.

The family-level classification of the genus *Itea* has been very complex, with the diversity of taxonomic relationships proposed by scientists reflecting the availability of new data sources. George Bentham correctly associated *Itea chinensis* with the Saxifragaceae family in his *Flora Hongkongensis*,[8] although he grouped it together with the distantly related genera *Dichroa* (now classified in the Hydrangeaceae[234]) and *Drosera* (the sundews, now classified in the Droseraceae[235]). Other taxonomists classified *Itea* in the Grossulariaceae (as in the current *Flora of Hong Kong*[233]) and Hydrangeaceae[74] based on comparative morphology. The availability of DNA sequencing technology in the 1990s, however, revealed a very unexpected association between several morphologically quite distinct families (collectively known as the 'Saxifragales') that had not previously been associated with each other.[33] Although molecular phylogenetic reconstructions have confirmed a relationship between *Itea* and the Grossulariaceae and Saxifragaceae (both core elements in the Saxifragales), *Itea* is now recognised as part of a separate evolutionary lineage, the family Iteaceae.[236,237] The novel hypotheses of taxonomic relationships that arose from the analysis of molecular data have acted as a significant stimulus for the reappraisal of patterns of morphological diversification in the group.

Plate 52. *Itea chinensis.* (A) Flowering branch, showing inflorescences composed of many small flowers. (B) Flowers. (C) Fruiting branch. (D) Single dehisced fruit capsule.

A

B

C

D

Kandelia obovata Sheue, H. Y. Liu & J. W. H. Yong

(= *Kandelia candel* auct. non (L.) Druce; *Kandelia rheedii* auct. non Wight & Arn.)

—— ✆ ——

Kandelia obovata (Kandelia; 秋茄樹 or 水筆仔; Rhizophoraceae)[129] is the most common tree species in Hong Kong's intertidal and estuarine mangroves. These mangroves are of great ecological significance: they not only provide a nursery habitat for an enormous diversity of species—including commercially important fish and crustaceans—but they also improve coastal water quality by removing agricultural pollutants and nutrients, and protect against coastal erosion.

The mangrove habitat is extremely challenging for plant survival, largely because of tidal inundations that cause major fluctuations in moisture and temperature; high salinity that results in physiological desiccation due to water loss; and oxygen deficiency due to anaerobic soils. *Kandelia obovata* possesses several adaptations to enable survival in these harsh conditions, including a thick outer waxy cuticle to reduce water loss, physiological mechanisms enabling salt tolerance,[238] and aerial roots that allow gaseous exchange.[239]

Several mangrove species, including *Kandelia obovata*, possess an unusual process known as 'vivipary' in which the seeds germinate precociously whilst still attached to the maternal plant.[130] The developing embryo penetrates through the fruit wall, emerging as an elongated 'hypocotyl' (the lower part of the immature shoot, below the 'cotyledons' or first-formed seed leaves): the structure that superficially appears to be an elongated fruit is therefore actually a young seedling, remaining attached to the maternal tree. Although some of these seedlings might conceivably fall in a vertical orientation and become embedded in the mud in close proximity to the maternal plant, it appears that most seedlings are horizontally oriented after falling and are carried further afield by water currents.[240] Once these horizontal seedlings have become rooted in mud they are able to reorient themselves vertically as a result of localised growth at the base of the hypocotyl, immediately above the roots; young seedlings of *Kandelia obovata* consequently often have a distinctive J-shaped hook.[131]

Plate 53. *Kandelia obovata.* (A) Flowering branch, showing axillary inflorescences of white flowers. (B) Flowers. (C) Seedling with elongated hypocotyl, remaining attached to the maternal plant and superficially resembling a fruit.

Leucaena leucocephala (Lam.) de Wit

(= *Leucaena glauca* (Willd.) Benth.)

—— ℰℐℴ ——

Leucaena leucocephala (White popinac; 銀合歡; Fabaceae)[45] is native to tropical America, but was an early introduction to Hong Kong, with records dating from at least 1860.[241] The species was subsequently included in the government's reafforestation programme in the 1920s and is now naturalised in many disturbed habitats.[25,242] Many different varieties are in cultivation globally, including some that form tall trees up to 15 m; those planted in Hong Kong are smaller (rarely exceeding 6 m) and less elegant, however, with a sparse crown.[80]

The leaves have a complex 'bipinnate' arrangement of leaflets ('pinnae') that are further subdivided into higher-order leaflets. The flowers are very small, with a reduced calyx of five fused sepals, and a corolla of five small petals; each flower is bisexual, with 10 stamens and a pistil that extend beyond the apex of the petals. The flowers are aggregated into spherical inflorescences up to 3 cm in diameter.

The fruits are typical of a member of the Fabaceae: the elongated pods (10–18 cm long) become dry and leathery at maturity, splitting open along opposite margins to release the seeds (6–25 per pod).[45] Although the seeds are apparently toxic to many animals, including horses and pigs,[79] it has nevertheless been suggested that the species is dispersed, at least in part, by cattle and goats, which have been reported to feed on the leaves and pods,[68] presumably transporting seeds in their dung.

A psyllid species (jumping plant louse belonging to the family Psyllidae) that is specific to *Leucaena leucocephala* first appeared in Hong Kong in the late 1980s.[51] This psyllid causes defoliation of the tree during the winter, but is ecologically valuable as a food source for various white-eyes, bulbuls and warblers—especially yellow-browed warblers, *Phylloscopus inornatus*, which overwinter amongst *Leucaena leucocephala* trees.[51]

Plate 54. *Leucaena leucocephala.* (A) Flowering branch, showing large compound leaves and spherical inflorescences of highly reduced flowers. (B) Single flower. (C) Mature, dry seed pods after they have split open to release their seeds.

Liquidambar formosana Hance

—— ⌘ ——

Liquidambar formosana (Sweet gum, Chinese Liquidambar or Formosan gum; 楓香; Altingiaceae)[200] is a deciduous native tree with characteristic three-lobed palmate leaves. Considerable evolutionary reduction is evident in the flowers consonant with wind pollination, including the loss of sepals and petals. The flowers are unisexual and aggregated into separate inflorescences: the comparatively small male inflorescences develop towards the end of the branches, whereas the larger, spherical female inflorescences occur towards the base. This spatial separation of the two sexes of flowers is an evolutionary adaption to reduce the chances of self-pollination and hence promote genetic diversity in the seeds produced.

The genus *Liquidambar* was formerly classified in the family Hamamelidaceae, although phylogenetic reconstruction based on comparisons of DNA sequence data now supports the recognition of the Altingiaceae as a separate family.[236]

Liquidambar is one of several genera that show a curious disjunct geographical distribution, split between East Asia and eastern North America. Interpretation of fossils and comparative analyses of DNA sequences have enabled an estimation of the timing of the evolutionary divergences that gave rise to these disjunctions. In most cases, the divergences occurred during the Miocene (23–5.3 million years ago),[243] demonstrating the influence of global climate change on plant distributions: global temperatures began to increase after the end of the Oligocene, peaking during the 'late Middle Miocene thermal maximum' (17–15 million years ago).[244] This resulted in the northward expansion of moist, warm temperate ecosystems, opening two main migration routes between east Asia and eastern North America: migration via the Bering Straits (with subsequent extinction in western North America); and migration via north Atlantic land bridges.[243] Molecular phylogenetic studies of *Liquidambar* suggest that the genus originated in Asia, with dispersal to North America via both migration routes.[6]

Plate 55. *Liquidambar formosana.* (A) Flowering branch, showing characteristic three-lobed palmate leaves and spherical female inflorescences. (B) Infructescence.

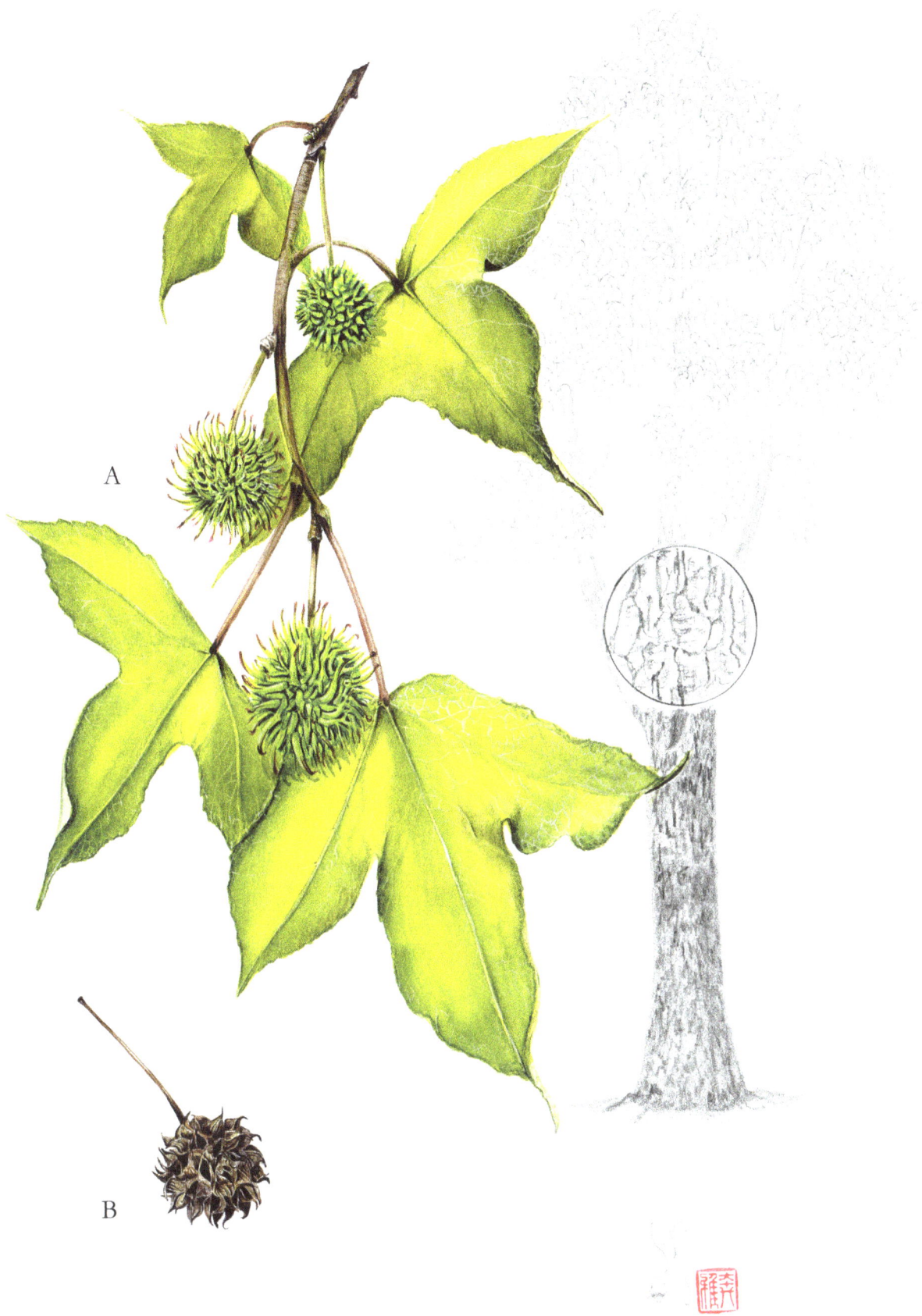

A

B

Lithocarpus glaber (Thunb.) Nakai
(= *Quercus thalassica* Hance)

—— ∾ ——

Lithocarpus glaber (Tanoak; 柯; Fagaceae)[142] is locally common in Hong Kong, growing in relict woodlands in remote ravines, and is occasionally planted.[1] It is closely related to *Cyclobalanopsis championii* (*q.v.*), although it is easily distinguished by the orientation of the male inflorescences, which are held upright in *Lithocarpus glaber* but pendent in *Cyclobalanopsis championii*.

Lithocarpus glaber flowers are commonly visited by calliphorid flies (and more rarely by syrphid flies) in Hong Kong.[61] The fossil record suggests that insect pollination is likely to be the ancestral mechanism in the oak family, Fagaceae,[245,246] with wind pollination (observed, for example, in *Cyclobalanopsis championii*, *q.v.*) derived independently in at least three lineages.[33]

The fruits of *Lithocarpus glaber* are acorns, in which a tough solitary nut is nested within a cup-like structure known as the 'cupule' that is inferred to have evolved from fused inflorescence branches.[143] As with many members of the Fagaceae, *Lithocarpus glaber* acorns are dispersed by 'scatter-hoarding' rodents such as squirrels, which cache the nuts in diverse locations with the intention of returning later to consume the seeds. Nuts that are forgotten or lost, and those hidden by rodents that die, are capable of germinating successfully. As noted elsewhere in this book, scatter-hoarding rodents are no longer represented in our local fauna,[139] and consequently tree species that rely on this dispersal mechanism are typically restricted to relict woodlands and are absent from secondary forests. Detailed ecological studies of seed predation and dispersal in a related species, *Lithocarpus harlandii*, have revealed direct evidence of this: although seed dispersal of populations in Sichuan in the presence of seed-caching rodents occurs over a distance of tens of meters,[247,248] seeds of the same species in Hong Kong merely accumulate at the base of each tree.[110]

Plate 56. *Lithocarpus glaber.* (A) Flowering branch, with upright male inflorescences. (B, C) Acorns, consisting of a nut nested within a basal cupule.

A

B

C

Litsea cubeba (Lour.) Pers.

(= *Litsea citrata* auct. non Blume; *Tetranthera polyantha* Wall.)

—— ☙ ——

Litsea cubeba (Fragrant Litsea or Mountain-pepper; 木薑子 or 山蒼樹; Lauraceae)[158] is locally common in Hong Kong, particularly in early stages of secondary forest succession.[1] The flowers have six 'tepals' (morphologically undifferentiated sepals and petals), arranged in two whorls of three; this contrasts strikingly with *Litsea glutinosa* (*q.v.*), which lacks a perianth entirely. The flowers are unisexual, with different sexes borne on different trees—a phenomenon known as 'dioecy,' which precludes self-fertilisation and hence promotes genetic diversity in the seeds produced. As is typical of other members of the Lauraceae family (*e.g.*, *Cinnamomum camphora* and *Machilus velutina*, *q.v.*), the functionally male flowers have nine fertile stamens in three whorls of three with paired nectar glands on the innermost whorl, and the functionally female flowers have a solitary carpel and nine sterile 'staminodes.' The flowers are predominantly visited by *Apis cerana* bees, but also by calliphorid and syrphid flies.[61]

The flowers are typically borne on the tree between late December and February, before leaf flushing occurs. In recent years, however, this temporal separation of flower and leaf formation appears to be less clear-cut, possibly because of global climate change: the tree used for the accompanying illustration, for example, was found to be in flower after the onset of leaf flushing. The flowering period of *Litsea cubeba* often coincides with the Chinese lunar New Year, and the pleasant fragrance of the flowers has previously resulted in indiscriminate cutting. *Litsea cubeba* was consequently one of the first species in Hong Kong to be awarded legal protection under the 1913 Licensing Ordinance, although it is no longer protected under the existing Forestry Regulations of the Forests and Countryside Ordinance.[2]

The small, black fruits are single-seeded 'drupes' (fleshy fruits with a tough inner fruit wall surrounding the seed). They are consumed by birds,[62] which are effective seed dispersers; the tough inner drupe wall protects the seed as it passes through the bird's digestive tract.

Plate 57. *Litsea cubeba.* (A, B) Flowering branch and inflorescence, showing white flowers with yellow stamens. (C) Cluster of fleshy drupes.

Litsea glutinosa (Lour.) C. B. Rob.

(= *Litsea sebifera* Pers.; *Tetranthera citrifolia* Juss.)

———— ❧ ————

Litsea glutinosa (Pond spice; 潺槁樹; Lauraceae)[158] is a medium-sized (to *c.* 15 m) evergreen tree, which is common and widespread in Southeast Asia. The species was first described by João de Loureiro in 1790 in his *Flora Cochinchinensis*[249] as '*Sebifera glutinosa.*' This rather gelatinous-sounding name is explained in Loureiro's text by the comment that the leaves and branches of the tree are soaked in water to extract a mucilage that is mixed with other ingredients to form a long-lasting plaster.[249] Other authors have highlighted similar traditional uses of *Litsea glutinosa* wood and bark infusions as a source of glue and as a hair pomade.[80,157]

The genus *Litsea* is very large, although estimates of species diversity vary greatly because of different opinions on its taxonomic delimitation.[250] The flowers are unisexual and are borne on separate individuals—a phenomenon known as 'dioecy,' which ensures that self-pollination and hence self-fertilisation cannot occur, thereby promoting the mixing of genes in the seeds produced. The flowers of most *Litsea* species possess a six-part perianth in two whorls of three, but this is either very highly reduced or entirely lacking in *Litsea glutinosa*. The reduced flowers are aggregated into small clusters that are surrounded by bracts that resemble a perianth: the inflorescences therefore superficially resemble flowers. The male flowers contain 15 or more stamens and a rudimentary pistil, and the female flowers contain a single pistil and numerous staminodes (sterile stamens).

Litsea glutinosa flowers are primarily visited by *Apis cerana* bees, but also by wasps, calliphorid flies, and hesperiid and lycaenid butterflies. Although bees are the most common floral visitor they may not be effective pollinators as most visits are to male trees to collect pollen and there is therefore little opportunity for pollen transfer to female trees.[61] The fruits are fleshy, *c.* 7 mm in diameter, and consumed by birds, which are effective seed dispersers.[62]

Plate 58. *Litsea glutinosa.* (A) Flowering branch, showing clusters of highly reduced flowers enclosed in perianth-like bracts. (B) Two inflorescence clusters. (C) Highly reduced male flower. (D) Cluster of fleshy fruits.

Livistona chinensis (Jacq.) R. Br. ex Mart.

——— ɕʓ ———

Livistona chinensis (Chinese fan palm or Fountain palm; 蒲葵; Arecaceae or Palmae)[251] is native to tropical southern China but is only cultivated in Hong Kong. It is a monocotyledon and hence belongs to an evolutionary lineage that had ancestrally lost the ability to develop wood. The palm-tree family (Arecaceae or Palmae) secondarily evolved an arborescent habit, although the mechanism responsible for the stem thickening differs from that evident in dicotyledons:[33] palm-tree wood is consequently structurally very distinctive, lacking the concentric rings that are visible in transverse sections of most other true wood.

The tree grows to 20 m in height and has a trunk that is marked with the scars of old leaves, with the petioles of fallen leaves remaining on the trunk for a considerable time. The leaves are very large (1–1.8 m in diameter) and fan-shaped, splitting into 20–30 cm-long terminal segments, each of which is folded. The lateral margins of the petioles are adorned with recurved spines (B in Plate 60).

The flowers are small (2–2.5 mm in diameter) and are aggregated into elongated inflorescences that are 1–1.5 m long. The calyx and corolla are both three-lobed, and each flower is bisexual, with six stamens and three carpels. Only one of the carpels in each flower is fertilised and hence the inflorescence develops into an elongated chain of separate fruits. The individual fruits (A in Plate 60) are blue-black and ellipsoidal (*c.* 1.8 × 1.2 cm), and are consumed by several different species of birds in Hong Kong, which presumably act as dispersal agents.[51] Although fruit bats have not been observed to eat the fruits locally, this has been reported from Peninsular Malaysia.[252]

Livistona chinensis trees serve as important roosting sites for Short-nosed fruit bats (*Cynopterus sphinx*): the bats employ 'tent-making' behaviour, modifying the palm fronds by chewing through the leaf veins to create roosting shelters.[51,253]

Plate 59. *Livistona chinensis* (fruits shown in Plate 60). Flowering individual, showing elongated inflorescences.

Plate 60. *Livistona chinensis* (flowering branch shown in Plate 59). (A) Fruits. (B) Older leaf.

A

B

Lophostemon confertus (R. Br.) Peter G. Wilson & J. T. Waterh.

(= *Tristania confertus* R. Br.)

——— ✃ ———

Lophostemon confertus (Brisbane box or Brush box; 紅膠木; Myrtaceae)[254] is native to subtropical eastern Australia, where it grows very rapidly and is drought-hardy and able to tolerate relatively poor soils.[80] These characteristics led to its inclusion in the extensive hillside reafforestation schemes undertaken in Hong Kong from the late nineteenth century.[25] The species not only grows well locally but flowers and sets seed abundantly, enabling locally produced seed to be sown; it consequently became one of the most commonly planted trees in Hong Kong, second only to *Pinus massoniana* (*q.v.*).[25] *Lophostemon confertus* is furthermore the source of a highly-prized timber that not only has a fine, attractive grain, but which is also resistant to many pests, including termites.[80] The financial benefits of cultivating these trees was undoubtedly one of the factors driving its inclusion in local reafforestation projects, although the rising costs of forestry operations and the availability of cheaper timber imports subsequently led to its decline.[25] *Lophostemon confertus* nevertheless continued to be widely planted in Hong Kong until the late 1990s;[25] the ecological value of the monoculture of introduced tree species has since been questioned, however, leading instead to the planting of native species alongside exotics.[255,256] Studies comparing levels of native plant colonisation in non-native tree plantations have revealed generally poor regeneration in sites dominated by *Lophostemon confertus*, although this might be due to local site conditions as tree species were selected for planting based on site characteristics.[53]

The flowers are bisexual and borne in clusters of 3–7, with five persistent sepals that are fused with the floral receptacle to form a cup-shaped 'hypanthium' characterstic of the Myrtaceae (*e.g.*, *Syzygium jambos*, *q.v.*), and five unfused white petals, *c.* 6 mm long. Each flower has numerous stamens that are fused into five bundles located immediately inside the petals, giving the flower a 'feathery' appearance. The pistil is derived from three fused carpels, and forms a dry, woody capsule (8–10 mm in diameter) that opens along three apical valves to release the seeds. The seeds are small (*c.* 2–3 mm long) and wind-dispersed.

Plate 61. *Lophostemon confertus*. (A) Flowering branch with clusters of white flowers. (B) Single flower, showing five white petals with clusters of fused stamens located immediately inside. (C) Isolated cluster of fused stamens. (D) Fruit capsules, showing apical valves open for the release of seeds.

Macaranga tanarius (L.) Müll. Arg.

———— ❧ ————

Macaranga tanarius (Elephant's ear; 血桐; Euphorbiaceae)[77] is a very common tree species and is widespread across Southeast Asia. It is a pioneer species, tolerant of salt and hence able to grow close to the shoreline. The English vernacular name is derived from the large 'peltate' leaves, in which the petiole (leaf stalk) is attached to the underside of the leaf lamina, some distance from the base of the lamina. The local Chinese name translates as 'bleeding tree,' in reference to the bright red sap that is evident when a branch is broken.

Many *Macaranga* species are closely associated with ants. Although *Macaranga tanarius* lacks the intimate mutualism exhibited by some species in the genus, in which ants are housed in specialised hollow twigs,[257] the species nevertheless has abundant extra-floral nectaries[49] that encourage ant populations, which in turn deter potential herbivores.[257,258]

The species is dioecious: the flowers are unisexual and are borne on separate individuals, thereby ensuring obligate cross-fertilisation and increasing levels of genetic variability in the offspring. The flowers are very highly reduced, lacking petals and with only a small calyx (1–2 mm long) and either 4–10 stamens (male flowers) or a single pistil derived from three fused carpels (female flowers).[77] The male flowers are borne behind green bracts in large, multi-flowered inflorescences, whereas the female flowers are borne in fewer-flowered inflorescences that grow in the axils of leaves.

An ecological study of a related *Macaranga* species has identified thrips (Thysanoptera) as pollinators, with the flowers providing a breeding site for the insects.[259] *Macaranga tanarius*, however, appears to show brood-site pollination by hemipteran 'flower bugs',[260] which are predators of thrips; it has therefore been conjectured that an evolutionary shift in pollination system may have occurred within *Macaranga*, with a change of pollinator from thrips to their predators.[260]

The fruits are rounded three-lobed capsules with soft spikes; the capsules split open at maturity to reveal black seeds that have a fleshy coat and are eaten by birds, which are the main seed dispersers for the species.[62]

Plate 62. *Macaranga tanarius.* (A) Fruiting branch, showing the characteristic peltate leaves. (B) Male inflorescence, with green bracts surrounding the highly reduced flowers.

A

B

Machilus chekiangensis S. K. Lee

(= *Machilus longipedunculata* S. K. Lee & F. N. Wei;
Persea longipedunculata (S. K. Lee & F. N. Wei) Kosterm.)

———— ❧ ————

Machilus chekiangensis (Chekiang Machilus or Zhejiang Machilus; 浙江潤楠; Lauraceae)[158] belongs to a species-rich genus (*c.* 100 species) that is well represented in the flora of Hong Kong. The delimitation of genera in the family Lauraceae has been contentious, with some taxonomists preferring to subsume *Machilus* within the genus *Persea*,[261] which includes Neotropical species such as the cultivated avocado, *Persea americana*. Recent research based on the evolutionary analysis of DNA sequence data supports the recognition of *Machilus* as a distinct genus,[262,263] however, and this is the approach adopted here and in the *Flora of Hong Kong*.[158]

Machilus chekiangensis is the commonest of Hong Kong's *Machilus* species and is often the dominant canopy-forming species. It is often confused with a superficially similar but rarer species, *Machilus thunbergii*, although the two species can be distinguished by reference to the floral perianth, which is significantly hairier in *Machilus chekiangensis*.

The small, pale yellow flowers are clustered in inflorescences that are borne at the base of branches derived from the current year's growth. Each flower is bisexual, with nine stamens in three whorls of three and a single carpel. The flowers are primarily visited by *Apis cerana* bees, but also by calliphorid and syrphid flies and pierid butterflies,[61] which are rewarded by nectar produced by the floral glands.

The flowers develop into small, spherical fruits (*c.* 7 mm in diameter) that are fleshy and turn bluish-black at maturity. The fruits are especially rich in fructose[63] and are accordingly attractive to birds,[62] which are responsible for dispersing the seeds. The fruits are borne on stalks that turn vivid red at maturity: this adaptation is presumably selectively advantageous because the contrasting colour of the stalk and the fruit is likely to render the ripe fruits much more visually conspicuous to birds.[264]

Plate 63. *Machilus chekiangensis.* (A) Flowering branch. (B) Flower, showing the six-part perianth. (C) Fruiting branch, showing the striking contrast in the colour of the fruits and fruit stalks.

Machilus velutina Champ. ex Benth.

(= *Persea velutina* (Champ. ex Benth.) Kosterm.)

—— ✐ ——

Machilus velutina (Woolly Machilus; 絨毛潤楠; Lauraceae)[158] is common in lowland secondary forests in Hong Kong. It can easily be distinguished from the many other local *Machilus* species (including *Machilus chekiangensis*, *q.v.*) as the lower leaf surface has a characteristically dense covering of rust-coloured hairs (B in Plate 64).

The Lauraceae represent one of the earliest surviving branches on the evolutionary tree of flowering plants and possesses many characteristics that have been interpreted as ancestral, including a poorly differentiated perianth. Since the sepals and petals are essentially morphologically indistinguishable, the term 'tepal' is often used.

The flowers are bisexual, with nine stamens in three whorls of three (the innermost of which are adorned with small stalked glands) and an additional inner whorl of sterile stamens ('staminodes') that are evolutionarily derived from stamens. Although the staminodes do not produce pollen, they are glandular and secrete nectar.[265] The flowers are therefore unusual in possessing two different sources of nectar: the stalked glands on the innermost functional stamens, and glands on the sterile staminodes. This can be explained by the fact that bisexual flowers in the Lauraceae exhibit two distinct sexual phases, with the carpel becoming receptive before the stamens release their pollen (a phenomenon known as 'protogyny,' which has evolved to prevent self-fertilisation within bisexual flowers). The glands on the staminodes secrete nectar during the female phase, whereas it is the stalked glands on the innermost stamens that produce nectar during the male phase.[265] The two glandular structures are significantly only capable of functioning once, as the process of secreting nectar involves the rupturing of the epidermis covering the gland.[265]

The flowers are primarily visited by *Apis cerana* bees, but also by wasps and calliphorid and syrphid flies[61] that are rewarded by the nectar. The small fleshy fruits are consumed by birds,[62] which are effective seed dispersers.

Plate 64. *Machilus velutina.* (A) Flowering branch, showing an inflorescence of small, white flowers. (B) Lower surface of a leaf, showing its rust-coloured hairs. (C) Fruits.

A

C

B

Magnolia championii Benth.

(= *Lirianthe championii* (Benth.) N. H. Xia & C. Y. Wu;
Magnolia pumila auct. non Andr.)

———— ⁊ ————

Magnolia championii (Hong Kong Magnolia; 香港木蘭; Magnoliaceae)[266] is a small tree (growing to 4 m) that is locally common in montane forests in Hong Kong.[1] It flowers in May and June, bearing small but highly fragrant flowers that are 1.5–2 cm in diameter. The sepals and petals are poorly differentiated and hence commonly referred to as 'tepals': the three outermost tepals are pale green (3.5–4 × *c.* 2 cm), whereas the inner tepals are whitish (2–2.5 × *c.* 1.5 cm).[266]

The Magnoliaceae were long considered archetypal 'primitive' flowering plants,[267,268] characterised by flowers with an elongated conical central axis, bearing numerous, spirally arranged unfused organs, and with little morphological differentiation between sepals and petals. The advent of molecular phylogenetics, in which evolutionary trees are reconstructed by comparing organisms' DNA sequences, has challenged these long-held beliefs, however. The monotypic genus *Amborella* (in the family Amborellaceae) is now recognised as a relict of the earliest diverging lineage of flowering plants, with the Magnoliaceae occupying a more derived position amongst early divergent angiosperms. Although *Amborella* exhibits many of the same floral characters as the Magnoliaceae (*e.g.*, spirally arranged organs of indeterminate number), these features are now inferred to be derived in the latter family.[33]

The separate carpels in the flower subsequently develop into separate (but closely appressed) fruits known as 'follicles.' This fruit type is also observed in *Illicium angustipetalum* (*q.v.*): as noted under the latter species, there is now some doubt whether follicles truly represent an archaic fruit type, with some researchers arguing that they are probably derived in the Magnoliaceae.[231,232] Each follicle splits open along the dorsal margin at maturity to expose its solitary scarlet seed, which hangs suspended on a delicate stalk derived from the spirally thickened xylem vessels that previously supplied water to the developing seed.[66]

Plate 65. *Magnolia championii.* (A, B) Flowering branches. (C) Fruit, composed of numerous follicles, opening to expose the red seeds that are initially suspended on thread-like stalks. (D) Seeds with attached stalks.

A

B

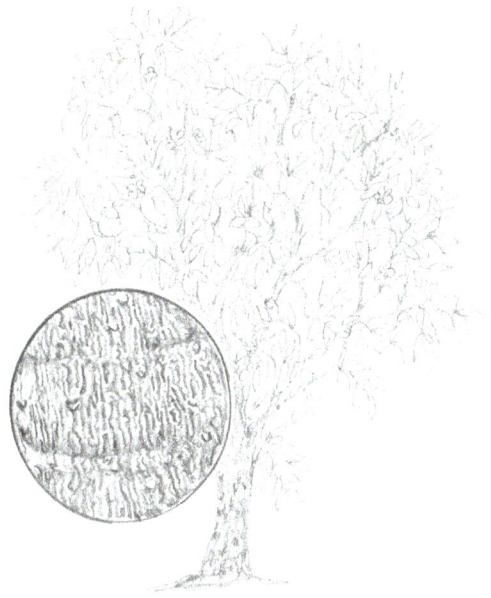

C

D

Mallotus paniculatus (Lam.) Müll. Arg.

(= Mallotus cochinchinensis Lour.; *Rottlera paniculata* (Lam.) A. Juss.)*

——— ☙ ———

Mallotus paniculatus (Turn-in-the-wind; 白楸; Euphorbiaceae)[77] belongs to the same taxonomic family as *Macaranga tanarius* (*q.v.*), with which it shares several important characteristics. The shape of the leaves is variable, ranging from ovate to distinctly three-lobed, and the petiole (leaf stalk) is attached to the lower surface of the leaf, slightly distant from the base of the leaf lamina. The lower surface of the leaf is densely covered in short white hairs and as a result is distinctly paler than the upper surface; movement of the leaves is very obvious in windy weather, hence explaining the English vernacular name. Each leaf has a pair of extra-floral nectaries at the top of the leaf stalk,[49,269] which presumably function in a similar way to those observed in *Macaranga*.

The flowers and fruits of *Mallotus paniculatus* are similar to those of *Macaranga tanarius*. As with *Macaranga*, individual trees are unisexual, bearing either male or female flowers, thereby ensuring cross-pollination and maximising genetic variability in the seeds produced. The flowers are borne in elongated terminal inflorescences and are highly reduced, lacking a perianth, with 50–60 stamens (male flowers) or a single pistil derived from three fused carpels (female flowers).[77] Although the flowers are commonly visited by bees, including *Apis cerana*, it appears that this is only to male plants, and it has been suggested that the species is more likely to be wind pollinated.[61]

Each fruit is covered with characteristic thick, soft spikes, and splits open at maturity to reveal three shiny back seeds that are dispersed by birds and squirrels.[62,270]

Plate 66. *Mallotus paniculatus.* (A) Flowering branch, showing elongated inflorescences. (B) Fruits with characteristic soft spikes, split open to reveal the black seeds.

A

B

Melaleuca cajuputi Roxb.
subsp. *cumingiana* (Turcz.) Barlow
(= *Melaleuca cumingiana* Turcz.; *Melaleuca leucadendra* auct. non (L.) L.;
Melaleuca quinquenervia auct. non (Cav.) S. T. Blake)

—— ❧ ——

Melaleuca cajuputi subsp. *cumingiana* (Paper-bark tree; 白千層; Myrtaceae)[254] is not native to Hong Kong but has been extensively planted in the territory as part of the government's reafforestation programme.[25] There is some evidence that the species is able to regenerate naturally in local forests,[51,53] although it has fortunately not become aggressively invasive as it has elsewhere.[271] The taxonomic identity of locally growing trees has been a source of considerable confusion, with the names *Melaleuca leucadendra* and *Melaleuca quinquenervia* erroneously used. Although it has been claimed that the species was introduced from Australia,[254] the subspecies that occurs in Hong Kong is actually native to Indochina, Peninsular Malaysia, Sumatra and Borneo.[272] The species has been extensively cultivated as a source of cajuput oil, which is extracted from the leaves and used medicinally as a decongestant.[273]

The English vernacular name describes the characteristic bark, which peels off in multiple paper-like layers (A, B in Plate 68). This complex bark structure may be an evolutionary adaptation to enable survival in areas that are prone to periodic fires by increasing flame resistance time and heat absorption, as has been shown for other species.[274]

The flowers are small, with a tubular calyx and five small, spreading petals. The stamens are *c.* 1 cm long and protrude from the flower, giving the inflorescence a feathery appearance. *Melaleuca cajuputi* superficially resembles Bottle-brush trees (*Callistemon* species) in the arrangement of its flowers, although it has five distinct bundles of fused stamens, a feature not observed in *Callistemon*; perhaps more conspicuously, however, the flowers of the locally cultivated *Callistemon* species are red, whereas those of *Melaleuca cajuputi* are white. The seeds, which are wind dispersed, develop in dry, woody fruit capsules (C in Plate 68) that remain on the branches for a considerable time. The prolonged protection of the seeds might also be an adaptation providing protection against fire.

Plate 67. *Melaleuca cajuputi* subsp. *cumingiana* (bark and fruit shown in Plate 68). (A) Flowering branch, showing inflorescences of flowers with protruding stamens. (B, C) Inflorescences at different developmental stages.

Plate 68. *Melaleuca cajuputi* subsp. *cumingiana* (flowering branch shown in Plate 67). (A) Bark, viewed from outside. (B) Bark, viewed from inside. (C) Fruiting branch.

A

B

C

Microcos nervosa (Lour.) S. Y. Hu

(= *Grewia microcos* L.; *Microcos paniculata* auct. non L.)

———— ✍ ————

Microcos nervosa (Microcos; 破布葉 or 布渣葉; Malvaceae)[275] is a small tree (growing to 12 m), native to southern China and common in lowland forests. The pale flowers develop in small clusters and have five distinct sepals (5–7 mm long) that alternate with an equal number of much smaller petals. Each flower is bisexual, with numerous stamens and a pistil derived from the fusion of three carpels. The flowers are primarily visited by bees, but also by butterflies.[276]

The fruits are small 'drupes' (berry-like fruits with a hard inner fruit wall surrounding the seeds), *c.* 7 × 10 mm, that become black and glossy at maturity. These drupes are eaten by birds and civets, which are effective seed dispersers.[40,62]

Extracts from the leaves of *Microcos nervosa* have been widely used as a traditional Chinese medicine as an analgesic[277] and for the prevention of coronary heart disease.[278] The species is now the focus of considerable pharmacological research.

The higher-level classification of the genus *Microcos* has been very contentious. Contemporary taxonomic research generally endeavours to ensure that all plant families represent distinct branches on the evolutionary tree (phylogeny): the circumscription of a family is accordingly changed when it is either shown to represent more than one branch in the phylogeny, or when it is shown to represent only part of a larger branch, with derived 'sub-branches' excluded from the family. *Microcos* was traditionally classified in the lime or linden family, Tiliaceae, but this family has been demonstrated to be highly heterogeneous, leading to the Tiliaceae being subsumed within an enlarged delimitation of the family Malvaceae.[279] Other researchers have objected to this, and have instead recognised several smaller families, with *Microcos* classified in the Sparrmanniaceae.[280] Biological classification is never stagnant: it changes as new data become available, reflecting the vibrant nature of such research.

Plate 69. *Microcos nervosa.* (A) Flowering branch, with an inflorescence of small flowers. (B) Detail of an inflorescence. (C) Single flower, with five conspicuous sepals. (D) Fruits.

B

C

A

D

Myrica rubra (Lour.) Sieb. & Zucc.

——— ∽ ———

Myrica rubra (Strawberry tree; 楊梅; Myricaceae)[281] is a small tree species (generally growing to *c*. 5 m in height, although sometimes up to 15 m) that is locally common in Hong Kong shrubland and secondary forest.[1] Its taxonomic affinities with oaks and their relatives (in the plant family Fagaceae) have long been recognised, with George Bentham, for example, classifying them together as the 'Amentaceae' in his *Flora Hongkongensis*.[8] This approach has received support from recent phylogenetic reconstructions based on DNA sequence data.[282] Many of the species in the lineage, including *Myrica rubra*[283] and *Casuarina equisetifolia* (*q.v.*), possess root nodules that maintain a symbiotic association with colonies of nitrogen-fixing filamentous bacteria, providing the trees with important nitrogen-rich nutrients and enabling the trees to thrive in nutrient-poor soils. There is evidence for multiple parallel evolutionary 'gains' of this symbiotic association in flowering plants, including the ancestor of the evolutionary lineage that includes both *Casuarina equisetifolia* and *Myrica rubra*.[33,284]

Myrica rubra flowers are highly reduced, lacking any sepals and petals but with 2–4 small bracts. Individual trees bear flowers of only one sex: male trees bear elongated catkin-like inflorescences (1–3 cm long), with flowers comprising 4–6 stamens with dark red pollen-bearing anthers; whereas female trees bear shorter inflorescences (0.5–1.5 cm long), with a pair of bright red divergent stigmas protruding from each flower.[281]

Myrica rubra has been cultivated for its fleshy edible fruits for over 2,000 years in China, although it is little known elsewhere.[285] The fruits are deep red or purplish-red 'drupes,' with a hard, inner fruit wall that protects the seed as it passes through the animal's gut after being swallowed. There are many reports of *Myrica* fruits being consumed by birds,[68] which presumably act as important seed dispersal agents, but there are also reports that *Myrica rubra* fruits are consumed by civets in Hong Kong.[62]

Plate 70. *Myrica rubra.* (A) Flowering branch of a female tree, with spikes of pistillate flowers showing pairs of bright red, divergent stigmas. (B) Male flower, showing stamens and bract. (C) Female flower, showing divergent stigmas and bracts. (D) Fruiting branch, showing deep red drupes.

A

B   C

D

Ormosia emarginata (Hook. & Arn.) Benth.

(= *Layia emarginata* Hook. & Arn.)

———— ⁶⁄⁹ ————

Ormosia emarginata (Emarginate-leaved Ormosia or Shrubby Ormosia; 凹葉紅豆; Fabaceae)[165] is a common local tree or shrub, growing up to 10 m in height. It was first collected by George Lay, the naturalist aboard *H. M. S. Blossom* during Captain Frederick W. Beechey's exploration of the Bering Strait and north Pacific Ocean (1825–28).[286] Although Lay never visited Hong Kong, the species was subsequently collected from Happy Valley between 1847 and 1850 by J. G. Champion,[165] a British army Captain and keen naturalist.

Ormosia species are characterised by their densely clustered compound leaves with 3–7 leaflets. *Ormosia emarginata* differs from the other local representatives of the genus by the notched apex to each leaflet. Although tropical and subtropical trees do not normally exhibit the prolonged leafless stage typical of temperate species, most have leaves that only last about one year; local tree species typically form new leaves between February and April, but *Ormosia emarginata* is unusual as it changes its leaves in mid-summer.[2]

Ormosia emarginata flowers are typical of the Fabaceae in possessing five clawed petals, comprising an enlarged upper petal (the 'standard', *c.* 7 × 8 mm), a pair of lateral 'wings' and two lower 'keel' petals. *Ormosia* flowers differ from most Fabaceae, however, since the 10 stamens are not fused. The flowers are pollinated locally by *Apis cerana* bees.[61]

The solitary carpel in the flower develops into a relatively short fruit pod (3–5.5 cm long), which splits open at maturity to expose up to four bright red seeds. These seeds superficially resemble fleshy berries: as with many other species in the Fabaceae (*e.g.*, *Adenanthera microsperma* and *Archidendron lucidum*, *q.v.*), this mimicry has evolved to deceive fruit-eating birds into dispersing the seeds without requiring the investment of energy-costly sugar rewards.[195]

Plate 71. *Ormosia emarginata.* (A) Flowering branch, showing morphologically distinct white petals. (B) Fruiting branches, with immature and mature pods; the latter have opened, exposing the bright red, berry-like seeds.

A

B

Osmanthus matsumuranus Hayata

——— ❧ ———

Osmanthus matsumuranus (Taiwan Osmanthus; 牛矢果; Oleaceae)[287] is locally rare and has only been recorded from a small number of localities in Hong Kong. Despite its English vernacular name, it is likely to be native to Hong Kong. It is a medium-sized tree (growing to 15 m), which bears branched inflorescences in the axils of the paired leaves. The inflorescences are composed of numerous small yellowish-white flowers that are variably bisexual or unisexual (in the latter case with flowers of the two sexes either borne on the same or different individual).[157] Each flower comprises a small four-lobed fused calyx, four fused petals (3–4 mm long), two stamens that are fused to the corolla tube, and a compound pistil; the stamens are sterile in functionally female flowers, and the pistil is correspondingly aborted in functionally male flowers.[287]

Osmanthus flowers are highly fragrant: the closely related species *Osmanthus fragrans* (which also grows in Hong Kong) is cultivated in China, with the flowers used for creating scented tea (after mixing with green or black tea leaves) and for flavouring desserts.[288] Although the floral scent undoubtedly functions as a pollinator attractant, there is also evidence that the scent contains volatile compounds that deter butterflies that would otherwise cause caterpillar damage to the foliage.[289]

The fruits are fleshy 'drupes' (15–25 × 8–12 mm), with a tough inner fruit wall layer surrounding the solitary seed, and turn purple-black when mature.[287] These features are likely to represent adaptations for seed dispersal by birds: frugivorous birds are often attracted to dark-coloured fruits, the drupes are an appropriate size for the bird's beak gape, and the tough inner fruit wall layer would protect the seed as it is passes through the bird's gut.

Plate 72. *Osmanthus matsumuranus.* (A) Flowering branch, with inflorescences of yellowish flowers. (B) Inflorescence. (C) Single flower, showing paired stamens with single pistil. (D) Fruiting branch, with purple-black drupes.

A

B

C

D

Paliurus ramosissimus (Lour.) Poir.
(= *Paliurus aubletia* Schult.)

———— ✍ ————

Paliurus ramosissimus (Thorny wing nut; 馬甲子; Rhamnaceae)[290] is locally common along forest margins and open coastal areas. The trees are small (growing to only 5 m in height), with a small spine growing at the base of each alternately arranged leaf. Each cream-coloured flower has five broadly triangular sepals that are basally fused to form a tube that surrounds the base of the pistil. The sepal lobes alternate with an equal number of much smaller, spoon-shaped petals (*c.* 1–1.5 mm long). The flowers are bisexual, with five stamens and a pistil derived from the fusion of three carpels.

The fruits are dry and hemispherical, with a distinctively flattened apex. This unusual fruit shape arises from the lateral swelling of the upper part of the carpel wall after fertilisation, forming a thickened wing across the top of the fruit.[291] Since the pistil comprises three fused carpels, the wing on the mature fruit is distinctively three-lobed and disk-like, and is conspicuously veined. In some species of *Paliurus* the fruit wing is very broad and thin and is likely to function in wind dispersal of seeds.[68] In *Paliurus ramosissimus*, however, the wing is narrower and is much thicker and wedge-shaped in longitudinal section—clearly poorly adapted for wind dispersal. It seems likely that *Paliurus ramosissimus* seeds are primarily water-dispersed: the species is commonly associated with coastal habitats (sometimes even being described as a 'semi-mangrove' species[292]), and the fruit wings are 'cork-like' and therefore able to float. Experiments to assess the buoyancy of *Paliurus ramosissimus* seeds in seawater have revealed that they remain afloat even after four months, with as much as 73% of the seeds successfully germinating despite prolonged exposure to saline conditions.[292] The widespread occurrence of *Paliurus ramosissimus* along forest margins in more inland sites is difficult to explain if the species relies on water dispersal: further research is necessary to clarify whether a secondary mode of dispersal exists.

Plate 73. *Paliurus ramosissimus.* (A) Flowering branch, showing small cream-coloured flowers. (B) Flower, with conspicuous sepals. (C) Fruiting branch, showing immature hemispherical winged fruits. (D) Mature fruits, with three-lobed wing.

Pandanus tectorius Parkinson

——— ❧ ———

Pandanus tectorius (Pandanus or Screw pine; 露兜樹; Pandanaceae)[293] is extensively cultivated in tropical regions and is common in Hong Kong, often growing near the coast. The trunk is branched, with conspicuous aerial roots and with spirally arranged leaves that are clustered at the apex of the stems. The leaves are characteristically spiny, with jagged barbs along the margins and the lower surface of the midrib (D in Plate 75).

Individual trees are unisexual, thereby preventing self-fertilisation and increasing the genetic diversity of progeny. Male individuals (Plate 74) bear large inflorescences composed of a terminal 'spadix' (up to 8 cm long) enclosed by a specialised bract (known as the 'spathe'). The inflorescences consist of vast numbers of very highly reduced male flowers, with each flower comprising 10–25 stamens that are basally fused into a column (C in Plate 74). The female flowers are aggregated in a terminal sub-spherical head, with each flower comprising 2–12 fused carpels. A perianth is absent in both the male and female flowers.

Although bees have been recorded visiting the male inflorescences and gathering pollen, they are apparently not attracted to the female inflorescences and therefore do not transfer pollen; the species is more likely to be wind pollinated given the vast quantity of pollen produced.[294] There is some evidence, however, of occasional asexual reproduction, in which fruit and seed development occurs without fertilisation of the ovule.[294]

The fruits of *Pandanus tectorius* are composed of aggregates of 40–80 fibrous 'drupes' (indehiscent fruits with a hard inner layer to the fruit wall). Each drupe (B, C in Plate 75) is composed of 2–12 fused segments, derived from the carpels of the female flower.[293] The fruits can remain buoyant and are commonly encountered in beach-drift; it is therefore likely that ocean currents may assist in long-distance dispersal.[294] There is also evidence from islands in the Pacific, however, of more localised dispersal by crabs[295] and flying foxes.[294]

Plate 74. *Pandanus tectorius* (fruits shown in Plate 75). (A) Flowering branch of male individual, showing inflorescence spadices with associated spathes. (B) Single male inflorescence spadix enclosed by a spathe. (C) Highly reduced male flower, composed of basally fused stamens. (D, E) Stamens.

Plate 75. *Pandanus tectorius* (flowers shown in Plate 74). (A) Fruiting branch, showing aggregated drupes. (B, C) Individual drupes. (D) Leaf, showing upwardly pointing marginal spines.

Pavetta hongkongensis Bremek.
(= *Pavetta indica* auct. non L.)

———— ❧ ————

Pavetta hongkongensis (Hong Kong Pavetta; 香港大沙葉 or 茜木; Rubiaceae)[67] is locally common in secondary forests in Hong Kong, including *fung shui* woods.[1] Individuals are often rather shrubby but can grow to *c.* 10 m. The leaves are characterised by conspicuous surface nodules that are typically more intensely pigmented than the surrounding leaf lamina. These nodules contain colonies of bacteria that are able to convert free atmospheric nitrogen into nitrogen-rich compounds that can be utilised by the tree.[296,297] The trees appear to be dependent on this symbiosis, with evidence for a complex mechanism to ensure that newly formed leaves are inoculated by the bacteria. The leaves are associated with 'stipules' (bract-like appendages at the base of the leaf stalks) that bear brush-like glandular hairs; these hairs produce a sticky exudate that harbours the bacteria, inoculating young developing leaves within the vegetative bud via their stomata.[296,298] There is also evidence for bacterial transmission between generations, with bacteria residing within seeds, adjacent to the embryo.[299]

Pavetta hongkongensis flowers are aggregated into loose inflorescences, and resemble those of *Adina pilulifera* (*q.v.*), which belongs to the same plant family, although *Pavetta hongkongensis* typically has floral organs in fours rathers than fives. The tiny sepals are fused into a small basal cup (*c.* 1 mm long), whereas the white petals are fused into a long tube (*c.* 15 mm long) with four outwardly projecting lobes. The stamens are located between the petal lobes at the mouth of the corolla tube, and each flower has an elongated style (*c.* 35 mm long) projecting far beyond the corolla.[67] The flowers are primarily visited by papilionid butterflies, but also by sphingid moths,[61] which probe into the flowers to collect nectar with their proboscis. The fruits that develop after the flowers have been fertilized are small fleshy berries (*c.* 6 mm in diameter), with the seeds likely to be dispersed by birds.

Pavetta hongkongensis is legally protected in Hong Kong under the Forestry Regulations of the Forests and Countryside Ordinance.[2]

Plate 76. *Pavetta hongkongensis.* (A) Flowering branch, with clusters of white, tubular flowers. (B, C) Leaf (upper surface) with its characteristic leaf nodules. (D) Part of an inflorescence with flowers with their protruding style. (E) Infructescence composed of small fleshy berries.

Pentaphylax euryoides Gardn. & Champ.

—— ❧ ——

Pentaphylax euryoides (Common Pentaphylax; 五列木; Pentaphylacaceae)[300] is a native species that is widely distributed in Southeast Asia and is very common in shrubland and young secondary forests in Hong Kong.[1] Individuals often have a rather shrubby habit and rarely exceed 10 m in height.

The small, white flowers are borne in axillary or terminal inflorescences, with the lowermost flowers opening first. Each flower is markedly pentamerous, with five rounded sepals (1.5–2.5 mm in diameter) and the same number of slightly elongated petals (4–5 mm long).[300] The flowers are bisexual, with five stamens alternating with the petals and with laterally expanded filaments forming a basal tube that is united with the petals, and a five-chambered carpel with five stigmas borne on a united style.

Pentaphylax euryoides flowers are pollinated by *Apis cerana* bees.[61] The resultant fruits are dry capsules that split along five valves at maturity to release the seeds, which are small (5–6 mm long) and winged, and dispersed by wind.[103]

The taxonomic affinities of *Pentaphylax euryoides*—the only species in the genus—have been the focus of considerable debate, with authorities often considering it a member of either the Theaceae or Ternstroemiaceae. Recent molecular phylogenetic research has supported its affinities with the Ternstroemiaceae,[71] although the latter family name has been displaced by the name Pentaphylaceae following inclusion of *Pentaphylax*. Two other representatives of the Pentaphylaceae are included in this book, *viz. Adinandra millettii* and *Anneslea fragrans*.

Plate 77. *Pentaphylax euryoides.* (A) Flowering branch. (B) Single flower, showing pentamerous arrangement of petals and stamens. (C) Fruiting branching, showing a capsule splitting open along five valves to release the wind-dispersed seeds.

Photinia benthamiana Hance

———— ✿ ————

Photinia benthamiana (Bentham's Photinia; 閩粵石楠; Rosaceae)[196] is locally common in secondary forests in Hong Kong.[1] It is a small tree, although individuals can reach up to 10 m in height. As is typical for the Rosaceae family, *Photinia benthamiana* flowers are bisexual and 'pentamerous' (with organs in fives or multiples of five): there are five small basally fused sepals, five white petals (3–5 mm long), 20 pink stamens, and a fused pistil.[196]

The position of the pistil relative to the other floral organs is highly variable in the Rosaceae, ranging from flowers with a 'superior' ovary that is located above the point of attachment of the sepals and petals, to others with an 'inferior' ovary that is embedded more deeply in the flower, below the sepals and petals. As discussed under *Anneslea fragrans* (*q.v.*), the evolutionary origin of the latter morphology endows a major selective advantage by protecting the ovary (and hence the ovules with their eggs) from floral visitors and potential herbivores.[81] Inferior ovaries have been identified as a key evolutionary innovation in many plant lineages[33] and their independent origins via different evolutionary routes is reflected by the diversity of floral morphologies. *Photinia* flowers, including *Photinia benthamiana*, have an inferior ovary that is surrounded by a 'hypanthium,' which is a cup-shaped extension of the floral receptacle, topped by the sepals and petals.[301]

As with many Rosaceae species, *Photinia benthamiana* forms fleshy fruits (known as 'pomes') in which the fruit wall is largely derived from the development of the hypanthium rather than the ovary wall. This fruit type will be most familiar to readers from commercially grown apples and pears, which characteristically retain their sepals at the apex of the fruit, on the opposite side from the fruit stalk; this feature is also observed in *Photinia benthamiana* and the related species *Eriobotrya japonica* (*q.v.*), which both have persistent sepals. Phylogenetic reconstructions have indicated that fleshy pomes are likely to have evolved from ancestors within the Rosaceae that had dry fruits.[302]

Plate 78. *Photinia benthamiana.* (A) Flowering branch, with clusters of small, white flowers. (B) Leaf venation. (C) Flowers. (D) Fruiting branch.

Phyllanthus emblica L.

——— ❧ ———

Phyllanthus emblica (Emblic or Myrobalan; 餘甘子; Phyllanthaceae)[77] is very common locally. The species superficially appears to have large, compound leaves consisting of small leaflets: closer examination reveals, however, that these 'leaflets' are actually small leaves (8–20 × 2–6 mm) that are alternately arranged on young branches.[77] The latter interpretation of morphology is clearly indicated by the presence of small but conspicuous scale-like appendages (known as 'stipules') at the base of the leaves: stipules are only ever associated with leaves, not with leaflets. The flowers are borne in the axils of these small leaves, giving the erroneous impression that they are borne within large compound leaves; this is presumably the origin of the generic name *Phyllanthus*, which literally translates as 'leaf-flower.'

The flowers are highly reduced and aggregated in small clusters that generally consist of a single female flower and 2–6 male flowers. Each flower has six sepals (1.2–2.5 mm long) but lack petals; the male flowers have three stamens, fused into a central column; and the female flowers have a pistil derived from the fusion of three carpels.[77]

Phyllanthus emblica fruits are small, rounded 'drupes' (fleshy fruits with a tough inner fruit wall surrounding the seed), and are rather acidic. They are primarily consumed by deer and colobine monkeys, although other animals such as squirrels and porcupines also eat them.[303] Deer apparently swallow entire fruits and then subsequently regurgitate the seed, still enclosed within the tough inner fruit wall; monkeys, in contrast, are less fastidious, taking a few bites from the fruit and dropping the remainder, including the seeds.[303] Muntjacs (barking deer) appear to be the most important seed dispersers in Hong Kong,[110] although it has also been suggested that civets are likely to eat the fruits.[62]

Phyllanthus emblica is widely used medicinally: the fruits have a very high vitamin C content and are used as an antiscorbutic and for treating various digestive ailments.[304] Fruit and leaf extracts have been shown to have anti-inflammatory and analgesic activity,[305,306] and the species is now the focus of considerable pharmacological research.

Plate 79. *Phyllanthus emblica.* (A) Flowering branch, with small alternately arranged leaves. (B) Female flower, with three stigmas. (C) Male flower, with three stamens fused to form a central column. (D) Fruiting branch.

A

B

C

D

Pinus elliottii Engelm.

———— ✤ ————

Pinus elliottii (Slash pine; 濕地松 or 愛氏松; Pinaceae)[307] is one of only two pine tree species growing in Hong Kong. It is native to south-eastern USA, but was extensively planted locally from the mid-1960s as part of the territory's reafforestation policy,[25] often alongside other exotics such as *Acacia confusa* and *Lophostemon confertus*. *Pinus elliottii* can easily be distinguished from the native pine species, *Pinus massoniana* (*q.v.*), as it has much larger and more robust leaves (18–25 cm long, compared with 10–18 cm) and larger cones (up to 14 cm long, compared with 4–7 cm).[307] The leaves—commonly known as 'needles'—occur in clusters of two or three.

Pine trees are gymnosperms, in which the reproductive organs are borne in cones rather than flowers. Each tree is bisexual, but with separate male and female cones; the male cones are small and borne in spiral clusters, whereas the female cones are considerably larger and solitary. The structure of the male cone is comparatively simple, consisting of membranous 'sporophylls' (fertile leaf-like structures) that bear pollen-containing sporangia. The larger female cones are structurally much more complex and consist of spirally arranged organs that bear the ovules; in terms of their evolutionary origin, each of these 'ovuliferous scales' is homologous with an entire male cone.[308]

All *Pinus* species are wind pollinated, with each pollen grain possessing two large air-filled cavities. These cavities possibly assist in keeping the pollen air-borne for longer, although recent research[309] has also suggested that they may also enable the pollen to float through the sticky 'pollination drop' that forms at the entrance to the downward-facing ovule, and hence assist in conveying the pollen closer to the egg. The reproductive cycle is very prolonged: although pollination occurs in the spring, the seeds do not mature until winter, one and a half years later. The female cones become increasingly woody as they develop and the component scales separate at maturity to release the seeds, which are winged to assist in dispersal.

Plate 80. *Pinus elliottii.* (A) Cone-bearing branch, showing long needles and small male cones. (B) Female cone. (C) Winged seeds.

A

B

C

Pinus massoniana Lamb.

(= *Pinus sinensis* Lamb.)

———— ⁊⁊ ————

Pinus massoniana (Chinese red pine; 馬尾松 or 山松; Pinaceae)[307] is a gymnosperm (cone-bearing plant), and the only pine species native to Hong Kong. There was very little forest in Hong Kong by the mid-nineteenth century, although *Pinus massoniana* was apparently maintained on lower hill slopes for firewood.[2,25] A major afforestation project was implemented locally from the early 1870s to meet growing demands for timber and firewood, peaking in the 1880s when a million or more trees were planted or seeds sown, the majority of which were *Pinus massoniana*.[2,25] Much of this forest was destroyed during the Second World War as demand for firewood increased following the interruption of fuel supplies from mainland China, with deforestation greatly accelerating during the Japanese occupation (1941–1945). Although subsequent reafforestation has taken place, this has been at a smaller scale.

The devastating effects of the pinewood nematode, *Bursaphelenchus xylophilus*, provide a cogent example of the ecological damage that can be caused by the introduction of an alien species. The pinewood nematode causes 'pine wilt disease' that can kill pine trees within six months of infection; the nematodes are spread by longicorn beetles (Cerambycidae), which have larvae that bore into the tree trunk.[3] The disease caused only limited damage in its native North America, but after its introduction to other parts of the world (including Hong Kong, where it was first confirmed in 1982), it caused widespread devastation of pine populations. In Hong Kong, *Pinus massoniana* proved to be more susceptible to the disease than other, non-native pines such as *Pinus elliottii* (*q.v.*).[2] The large-scale loss of *Pinus massoniana* has inevitably had a major impact on the local landscape, but that loss has subsequently been partially countered by aerial seeding.

Plate 81. *Pinus massoniana.* (A) Cone-bearing branch, showing cluster of male cones. (B, C) Female cones (closed and open).

A

B

C

Podocarpus macrophyllus (Thunb.) Sweet

———— ❧ ————

Podocarpus macrophyllus (Buddhist pine or Kusamaki; 羅漢松; Podocarpaceae)[310] is a medium-sized evergreen tree (growing to 15 m) with stiff, strap-shaped leaves. It has a restricted distribution in Hong Kong[136] but has been extensively cultivated locally as an ornamental.[80] The trees are highly prized for *fung shui* purposes and are therefore very valuable: there have been several reports in the media of local trees being illegally uprooted and smuggled into mainland China.

Podocarpus macrophyllus is a gymnosperm (literally 'naked-seeded plant'), in which the ovule—which contains the egg cell—is not surrounded by a protective ovary wall as in flowering plants. Populations consist of separate male and female trees. Male individuals bear catkin-like cones in clusters of 3–5, composed of numerous, spirally arranged scales, each of which bears two pollen sacs.[311] The pollen grains have a pair of air bladders that assist in wind dispersal.

The female reproductive structures are small, solitary, and borne in leaf axils; they consist of a single ovule on a short stalk attached to the upper surface of a bract. Each ovule is recurved so that the opening at the apex of the ovule is towards the base of the bract. A sticky 'pollination drop' forms near this opening, trapping the wind-borne pollen; the air bladders on the pollen grains act as a flotation device, helping move the pollen upwards towards the aperture of the ovule.[311-313]

Since *Podocarpus macrophyllus* is a gymnosperm, the seeds are borne directly on the branch and are not protected by a fruit wall as in flowering plants. The mature seeds have a purplish-black 'epimatium' (a swollen appendage derived from the ovuliferous bract) and the ovule stalk becomes fleshy and turns bright red at maturity. These fleshy seed stalks are attractive to birds, which are effective seed dispersers for the species.[314] The seed stalk therefore performs a similar function to the fleshy fruit coat in many flowering plants, although it is anatomically quite different and has an independent evolutionary origin.

Plate 82. *Podocarpus macrophyllus.* (A) Male reproductive shoot, bearing catkin-like male cones. (B) Female reproductive shoot, bearing small solitary ovules. (C) Female reproductive shoot with seeds. (D) Single seed with purplish-black epimatium, borne on a fleshy red stalk that superficially resembles a fruit.

A

B

C

D

Polyspora axillaris (Roxb. ex Ker Gawl.) Sweet

(= *Camellia axillaris* Roxb. ex Ker Gawl.;
Gordonia anomala Spreng.; *Gordonia axillaris* (Roxb. ex Ker Gawl.) D. Dietr.)

——— ✃ ———

Polyspora axillaris (Gordonia or False Camellia; 大頭茶; Theaceae)[69] is a common evergreen tree in Hong Kong. For many years the species was treated as a member of the genus *Gordonia*; reconstructions of evolutionary relationships based on the analysis of DNA sequence data, however, have revealed that the Chinese species assigned to *Gordonia* are not closely related to the North American species for which the generic name had first been used, and that the name *Polyspora* should be adopted instead.[5]

The flowers are very conspicuous, with large, showy white petals (3.5–5 cm long). They are primarily pollinated by large vespid wasps, although a diversity of other animals also visit the flowers, including bees, butterflies, moths and birds.[61,315] Papilionid butterflies are commonly observed to feed off the nectar, and although they sometimes collect pollen grains on their body they rarely contact the stigma and are therefore primarily 'nectar thieves' rather than pollinators.[315] Flowering occurs locally between September and December, but the fruits do not reach maturity until the following winter,[195] between October and April.

The fruits mature as dry capsules that split open apically to release winged seeds that are wind dispersed and hence disseminated over considerable distances. *Polyspora axillaris* is accordingly a pioneer, often colonising previously disturbed areas where animal-dispersed species are rare. It thrives in nutrient-poor soils and is often dominant in shrubland and during early stages of secondary forest succession.[316]

Plate 83. *Polyspora axillaris.* (A) Flowering branch, showing conspicuous flowers. (B) Individual flower. (C) Carpel. (D) Stamen. (E) Fruits.

A

E   D

C   B

Pyrenaria spectabilis (Champ.) C. Y. Wu & S. X. Yang

(= *Camellia reticulata* auct. non Lindl.; *Camellia spectabilis* Champ.;
Tutcheria spectabilis (Champ.) Dunn)

——— ✑ ———

Pyrenaria spectabilis (Common Tutcheria; 石筆木; Theaceae)[69] is a rare local species that is now legally protected under the Forestry Regulations of the Hong Kong Government's Forests and Countryside Ordinance.[98] Although attempts to include the species in official reafforestation programmes during the first half of the twentieth century were unsuccessful,[25] recent research has shown that the species has some potential for forest restoration on degraded hillsides in Hong Kong.[317]

The flowers of *Pyrenaria spectabilis* are comparatively large and showy (4–7 cm in diameter), with 9–11 rounded and overlapping sepals, five or six conspicuous white petals (4–5 × 2–3 cm), numerous stamens that are fused to the base of the petals, and a compound pistil with its prominent lobed stigma projecting out of the flower to receive pollen.

Pyrenaria spectabilis fruits superficially resemble those of species in the closely related genus *Camellia* (*e.g.*, *Camellia crapnelliana* and *Camellia hongkongensis*, *q.v.*), but can be distinguished by the direction in which the fruit opens (splitting from the base in *Pyrenaria*, but from the apex in *Camellia*).[69] Although little is known of the seed dispersal mechanism, it seems likely that *Pyrenaria spectabilis* is another example of a species that is dispersed by 'scatter-hoarding' rodents that collect and store fruits for later consumption; seeds can successfully germinate when the rodent fails to retrieve the fruit. Significantly, however, rodents capable of effecting this dispersal no longer survive in Hong Kong,[139] and as a consequence, species with this dispersal mechanism are rare and noticeably absent from secondary forests. This is in striking contrast with the closely related species *Polyspora axillaris* (*q.v.*), which produces winged seeds that are wind-dispersed: *Polyspora axillaris* is far more common in Hong Kong, and is regarded as a pioneer species that is able to colonise disturbed area and often dominates early-stage secondary forests.[316]

Plate 84. *Pyrenaria spectabilis.* (A) Flowering branch, showing large showy flowers. (B) Fruiting branch, with dry fruit capsules. (C) Single fruit.

A

B

C

Pyrus calleryana Decne.

———— ✦ ————

Pyrus calleryana (Callery pear or Wild pear; 豆梨 or 麻子梨; Rosaceae)[196] is native to southern China but is comparatively rare in Hong Kong.[1] It grows to *c.* 8 m in height and is characterised by clusters of attractive white flowers with five distinctly clawed petals (*c.* 13 × 10 mm) and numerous conspicuous stamens. The fleshy fruits are greenish-brown and *c.* 1 cm in diameter.

Pyrus species have been cultivated as fruit trees for thousands of years, with the commercial pear obtained from selected varieties of *Pyrus communis*. The United States Department of Agriculture funded an expedition to China in 1917 in search of wild pear seeds, with the objective of identifying disease-resistant genes that could be used for breeding programmes involving cultivated varieties.[318] *Pyrus calleryana*—which was first described in the nineteenth century[319] and introduced into cultivation in 1908 as an ornamental tree[318]—was one of the major sources of these seeds.

Many *Pyrus* species, including *Pyrus calleryana*, can interbreed freely and form fertile hybrids. In some cases, fruitset is enhanced if different pear 'cultivars' (varieties artificially produced and maintained in cultivation) are planted in close proximity to encourage such hybridisation.[318] *Pyrus calleryana* cultivation was heavily promoted in the United States from the 1950s and 1960s, leading to its widespread adoption as an ornamental tree. Concerns regarding its potential as an invasive weed began to be expressed in the 1990s, however, with the realisation that it was not only flowering profusely and able to set seed within only three years, but was also hybridising extensively.[318] The propagation of *Pyrus calleryana* trees from many different source populations and the consequent hybridisation between these cultivars significantly promoted genetic diversity in the resultant seeds and acted as a major stimulus for its invasive spread: unlike many other weeds, the invasiveness of *Pyrus calleryana* is not due to any 'fitness' advantage inherent in its hybrid origin.[320,321] Hybridisation—irrespective of whether it is between different species or between genetically distinct lineages within a species—is now recognised as an evolutionary pathway that can significantly promote invasive potential.[322]

Plate 85. *Pyrus calleryana*. (A) Flowering branch, showing clusters of white flowers. (B) Single flower. (C) Cluster of fruits ('pomes').

A

B

C

Reevesia thyrsoidea Lindl.

———— ❧ ————

Reevesia thyrsoidea (Reevesia; 梭羅樹; Malvaceae)[213] is a common evergreen tree, native to southern China, with leaves characteristically crowded at the ends of branches. The flowers are small, with five spreading white petals, but are aggregated into large rounded inflorescences. The stamens in each flower are fused into an almost spherical mass borne at the apex of a conspicuous tubular extension of the floral receptacle known as the 'androgynophore,' similar to that observed in *Sterculia lanceolata* (*q.v.*), which belongs to the same plant family. The pollen sacs of the fused stamens are irregularly arranged on the outer surface of the spherical head. The carpels are held within the fused stamen mass and are also borne on the androgynophore.

The flowers are primarily pollinated by butterflies, but are also visited by diurnal hawkmoths and nocturnal moths.[61] The insects brush against the pollen sacs and deposit pollen on the stigmas as they probe into the flower with their proboscis to collect nectar. The carpels develop into dry, pentamerous fruit capsules at maturity. The fruits split open to release the seeds, which have a wing at one end, and are wind-dispersed.

The species and genus were first described in 1827 by John Lindley[323] based on specimens collected by John Reeves (1774–1856) in southern China. Reeves was a keen amateur naturalist who worked as a tea inspector in Canton (now known as Guangzhou) from 1812 until 1831.[324] As well as making the first collection of *Reevesia thyrsoidea*, Reeves was responsible for amassing an enormous collection of plant and animal specimens and natural history paintings that are now held in the Natural History Museum in London.

Plate 86. *Reevesia thyrsoidea.* (A) Flowering branch, showing large inflorescences composed of small white flowers with distinctive central 'androgynophore.' (B–D) Fruit capsules. (E) Winged seed.

Rhaphiolepis indica (L.) Lindl. ex Ker

(= *Crataegus indica* L.)

——— ❧ ———

Rhaphiolepis indica (Hong Kong hawthorn; 石斑木, 車輪梅 or 春花; Rosaceae)[196] is a shrubby species (rarely growing as a small tree to *c.* 4 m in height) that is very common in shrublands, grasslands and early-stage secondary forests, especially near streams.[1] It is a light-demanding species that is well adapted for colonising open habitats as its seeds are typically bird-dispersed. Another explanation for the widespread occurrence of *Rhaphiolepis indica* on Hong Kong's hillsides is its ability to regenerate well after fire.[325] Hill fires are very common, especially when traditional grave-sweeping ceremonies coincide with dry weather: it has been estimated that there are around 300 outbreaks annually and that over half of the land within Hong Kong's country parks is burnt every decade.[326]

Rhaphiolepis indica flowers resemble the other representatives of the Rosaceae family included in this book (*Eriobotrya japonica*, *Photinia benthamiana* and *Pyrus calleryana*, *q.v.*) in having five fused sepals and five pinkish-white petals (5–7 × 4–5 mm), with both whorls attached to an extension of the floral receptacle known as the 'hypanthium.'[196] The flowers are bisexual, with 15 stamens and a 2–3-chambered pistil, and are fragrant, attracting a broad range of insects including bees (*Apis cerana*), and nymphalid, papilionid and pierid butterflies.[61]

As described for the other Rosaceae species illustrated in this book, the fruits are small 'pomes,' in which the flesh is derived from growth of the hypanthium rather than the ovary wall. The fruit flesh is sugar-rich, with high levels of glucose and moderate levels of fructose, but with no detectable sucrose.[63] This is typical for fruits consumed by birds, since birds typically do not possess the sucrase enzyme that is required to break down sucrose. Not surprisingly, the fruits are reported to be eaten by birds,[62,63] although there are also reports of seed dispersal by civets.[63]

Plate 87. *Rhaphiolepis indica.* (A) Flowering branch, with clusters of flowers. (B, C) Flowers with pinkish-white petals. (D) Fruiting branch, showing clusters of small, purplish-black 'pomes.'

Rhodoleia championii Hook.

———— ❧ ————

Rhodoleia championii (Rhodoleia; 紅花荷, 紅苞木 or 吊鐘王; Hamamelidaceae)[200] bears small pendent inflorescences that superficially resemble individual flowers but are actually clusters of 5–8 highly reduced flowers, with each inflorescence surrounded by multiple sepal-like involucral bracts. Each flower within the inflorescence is bilaterally symmetrical, with a rudimentary calyx and a conspicuous corolla consisting of 1–4 petals (occasionally absent) that develops on the abaxial side of the flower, towards the outside of the inflorescence.

Unlike most species in the Hamamelidaceae, which have unisexual flowers that are small and inconspicuous, *Rhodoleia championii* has bisexual flowers that are vividly coloured. They are primarily pollinated by birds, with well-documented local records of visits by Japanese white-eyes (*Zosterops japonica*) and Fork-tailed sunbirds (*Aethopyga christinae*).[61,327] The flowers illustrate many of the characteristic features typically associated with bird pollination, including: red pigmentation, absence of scent, copious nectar production, and robust floral structures that are able to withstand vigorous probing by birds.[327] The seeds are winged and are wind dispersed.

Rhodoleia championii was first collected in 1849 by J. G. Champion, a British army Captain and keen naturalist who made extensive plant collections in Hong Kong between 1847 and 1850. Champion discovered the species in a forested hillside near 'Little Hong Kong,' an area close to Wong Chuk Hang that was destroyed in the 1970s during the development of the Aberdeen Tunnel. William J. Hooker, the Director of Kew Gardens, named the species in Champion's honour in 1850.[328] *Rhodoleia championii* is now legally protected under the Forestry Regulations of the Hong Kong Government's Forests and Countryside Ordinance.[98]

Plate 88. *Rhodoleia championii.* (A) Flowering branch, showing flower-like inflorescences. (B) Large petal. (C) Stamen. (D) Pistil derived from two fused carpels, with two styles. (E) Cluster of fruits (infructescence).

A

B

C

D

E

Rhus chinensis Mill.

———— ✃ ————

Rhus chinensis (Sumac; 鹽膚木; Anacardiaceae)[329] is a common tree in shrubland and forest margins. The leaves are subdivided into 7–13 small leaflets with very distinctive lateral wings on the 'rachis' (the central axis of the leaf), between leaflets. These leaves often possess galls that are induced by an aphid, *Schlechtendalia chinensis*.[330,331]

The sap of *Rhus chinensis* contains an allergen that can cause dermatitis.[332] As is often the case with poisonous plants, however, *Rhus chinensis* is also of considerable medicinal importance, with the leaves, roots, stem, bark and fruit of the species having been shown to have various antiviral, antibacterial and anticancer applications.[332,333] The leaf galls created by the aphids are also widely used and are the source of a medicine known as '*Galla chinensis*' or 五倍子. Gall formation is encouraged by enclosing the aphids in bags over the leaves; the resultant galls are then harvested before they split open and are immersed in boiling water to kill the aphids, before the galls are air-dried.[332]

The flowers are small (with white petals *c.* 2 mm long) but are aggregated into large inflorescences that are 20–30 cm long.[329] Individual flowers are either bisexual or functionally unisexual—in the latter case either co-occurring on the same tree (monoecy) or on separate trees (dioecy)[157]—and are pollinated by bees.[52]

After fertilisation, the flowers develop into small dry single-seeded fruits; the skin of the fruit is red and densely covered in short hairs.[329] The fruits are consumed by birds, which are responsible for dispersing the seeds.[40,334]

Plate 89. *Rhus chinensis.* (A) Flowering branch, showing distinctive lateral wings on the leaf rachis between leaflets. (B) Inflorescence. (C) Isolated flower. (D) Infructescence.

—201—

Sapium discolor (Champ. ex Benth.) Müll. Arg.

(= *Stillingia discolor* Champ. ex Benth.; *Triadica cochinchinensis* Lour.)

———— ∾ ————

Sapium discolor (Mountain tallow; 山烏桕; Euphorbiaceae)[77] is a distinctive species with red pigmentation of young leaves that often persists into maturity, especially along the midrib and leaf margins; this is presumably the origin of the specific epithet '*discolor*,' which means 'differently coloured.' As with other species in the Euphorbiaceae (*e.g.*, *Aleurites moluccana*, *Endospermum chinense*, *Macaranga tanarius*, *Mallotus paniculatus* and *Vernicia montana*, *q.v.*), the leaves possess a pair of extra-floral nectaries at the base of the leaf lamina, which produce a sugary exudate that promotes ant colonies that in turn help to deter herbivores.[49]

The species is very common in secondary forests[1] and is a pioneer species that often dominates forests during early stages of ecological succession.[2] It is very light demanding, and is eventually out-competed by more shade-tolerant species.[316]

Sapium discolor flowers are unisexual and highly reduced, lacking petals and with three very small sepals: the male flowers have 2–3 stamens; and the female flowers have a fused pistil with three stigmas. The flowers are aggregated into elongated inflorescences that are 4–9 cm long, with several solitary female flowers at the base and numerous male flowers on the upper part.[77] The flowers are primarily visited by *Apis cerana* bees, but also by wasps.[61]

Sapium discolor is deciduous, and is leafless in Hong Kong between December and March; ripe fruit are borne when the leaves are turning red before being shed, and this may act as a visual cue to help attract frugivores.[2] The fruits have a high fat content (70% of dry weight),[2] providing a good energy source for many different species of frugivorous birds.[40,62]

The nomenclature of this species is controversial, with some authors believing that the name *Sapium discolor* is a later synonym of *Triadica cochinchinensis*.[335] We use the name *Sapium discolor* to be consistent with the treatment in the *Flora of Hong Kong*.[77]

Plate 90. *Sapium discolor.* (A) Flowering branch, showing elongated inflorescence composed of highly reduced flowers. (B) Cluster of female flowers after fertilisation. (C) Immature and mature (dehiscent) fruits.

B

A

C

Sarauia tristyla DC.

——— ✑ ———

Sarauia tristyla (Sarauia; 水東哥 or 米花樹; Actinidiaceae)[336] is a small native tree species (growing to *c.* 5 m) that is commonly encountered along streams. The flowers, which are borne in branched inflorescences, have five small sepals (3–4 mm long) and five pinkish-white petals (*c.* 8 mm long) that are splayed outwards towards the apex. Each flower is bisexual, with numerous stamens attached to the base of the petals and a fused carpel with 3–5 chambers, each containing numerous ovules.[336] Although the specific epithet '*tristyla*' indicates that the flowers have three styles (and hence three stigmas), there is actually some variation, with either three or four (or rarely five) styles.

Studies of other *Sarauia* species have indicated that the flowers are bee-pollinated.[337-339] Although *Sarauia tristyla* flowers are bisexual, many species in the genus exhibit a degree of sexual dimorphism that promotes cross-pollination and genetic mixing in the resultant seeds. In the sexually dimorphic species, individual trees bear a combination of flowers that are functionally bisexual and others that are functionally male (with functional stamens and a sterile, vestigial carpel)—a rare phenomenon known as 'androdioecy.'[338] Other studies have suggested that some of these androdioecious species may have functionally distinct male and female flowers, however, with structurally bisexual flowers producing sterile pollen; this sterile pollen serves as a reward to pollen-collecting bees that would otherwise not be attracted to female flowers.[339]

Sarauia tristyla fruits are small, cream-coloured berries (6–10 mm in diameter) with a succulent flesh. Although there are no specific studies of frugivory of this species in Hong Kong, studies from Thailand reveal that the berries are consumed by bulbuls, gibbons and macaques, which serve as effective seed dispersers.[270]

Plate 91. *Sarauia tristyla.* (A) Flowering branch with multi-flowered inforescences. (B) Single flower, with five recurved pink petals. (C) Globose, cream-coloured berries.

Schefflera heptaphylla (L.) Frodin

(= *Schefflera octophylla* (Lour.) Harms)

Schefflera heptaphylla (Ivy tree, Goosefoot tree or Duckfoot tree; 鴨腳木 or 鵝掌柴; Araliaceae)[340] is widespread in east Asia and is one of the commonest trees in secondary forests in Hong Kong. It is easily recognisable by its distinctive palmately compound leaves, composed of 5–10 leaflets; this variability in leaflet number has resulted in the superficially contradictory Latin epithets '*heptaphylla*' and '*octophylla*' that have previously been adopted for this species.

The flowers are fragrant and produce copious nectar, and are visited by diversity of bees, wasps, flies and butterflies.[61,341] Although the flowers are very small (approximately 5 mm in diameter), they are aggregated in large inflorescences, up to 25 cm long. Each tree typically bears 10–15 inflorescences that are composed of up to three orders of smaller clusters; it has been estimated that each inflorescence consists of 1250–1400 flowers and hence each tree produces tens of thousands of flowers each season.[341] Maturation of the inflorescences and their constituent clusters is sequential, thereby greatly extending the flowering period of each tree. Individual flowers are bisexual (bearing both stamens and carpels), but the stamens mature first, consequently preventing self-pollination within the flower and hence increasing levels of genetic diversity in the seeds produced.[341]

Schefflera heptaphylla flowers in winter (typically from November until the following January locally), with fruits (small berries, 3–4 mm diameter) ripening between February and March. As the flowers and fruits form during winter, the trees are ecologically very important as a food source for insects (which consume the nectar produced by the flowers[61,341]) and birds and fruit bats (which consume the berries[40,62,163]) during periods when food is comparatively scarce.

Plate 92. *Schefflera heptaphylla.* (A) Flowering branch, showing the characteristic palmately compound leaf, and the large compound inflorescence composed of numerous small flowers. (B) Small cluster of flowers within elongate inflorescence. (C) Individual flower. (D) Stamen. (E) Fruits.

Schima superba Gardner & Champ.

(= *Schima noronhae* auct. non Reinw. ex Blume)

———— ☙ ————

Schima superba (Schima or Chinese guger tree; 木荷 or 荷樹; Theaceae)[69] is locally common in Hong Kong, including *fung shui* woods. The species is known to be indigenous to the territory, with seeds recovered during an archaeological excavation of a 6,000-year-old site at Penny's Bay in Lantau.[2] *Schima superba* is now one of the most extensively planted of our native species, having been successfully grown in established pine plantations over the past half-century.[25]

Schima superba was first described posthumously by George Gardner in 1849 from specimens collected by J. G. Champion, a British army Captain and keen botanist, from 'Wingnychery' (Wong Nai Chung), Hong Kong Island.[342] Gardner applied the specific epithet '*superba*' in recognition of the impressive floral display when trees are in full bloom between June and August.

The flowers are bisexual, with five small sepals, five conspicuous white petals (1–1.5 cm long), numerous yellow stamens, and a pistil derived from the fusion of five carpels. The flowers are highly fragrant and are primarily pollinated by *Apis cerana* bees and pierid butterflies, although they are also visited by wasps, papilionid and nymphalid butterflies, as well as sphingid moths.[61]

The flowers develop into woody 'capsules' (1–1.5 cm in diameter) that split open apically into five equal segments, releasing the seeds that are narrowly winged and dispersed by wind.[103,163] Squirrels are often attracted to *Schima superba* fruits, although they are likely to be seed predators rather than seed dispersers since they are observed to break open immature capsules to eat seeds.[2]

Plate 93. *Schima superba.* (A) Flowering branch, bearing conspicuous white flowers. (B) Flower after removal of petals. (C) Compound pistil. (D) Fruit capsule and winged seeds.

Scolopia chinensis (Lour.) Clos

(= Phoberos chinensis Lour.; *Scolopia crenata* auct. non Clos)*

———— ℰℐ ————

Scolopia chinensis (Chinese Scolopia; 刺柊; Salicaceae)[219] is a native tree species that is locally common in *fung shui* woods and in coastal secondary forests.[1] Although it is not a 'true' mangrove species, *Scolopia* species are often regarded as 'mangrove associates' and are presumably resistant to periodic saline inundations. *Scolopia chinensis* has characteristic stout spines on the trunk and branches, and elliptic leaves that are adorned with a pair of glands on opposite sides of the petiole, close to the leaf lamina.

The flowers are bisexual, with 4–6 small sepals (1–2 mm long) and an equal number of slightly longer pale-yellow petals. Each flower has numerous elongated stamens with a hairy apex that extends beyong the pollen-bearing anther, and a single-chambered pistil surrounded by a 10-lobed fleshy glandular disk.[219]

The fruits are round red berries (*c.* 4 mm in diameter) with a remnant of the style persisting at the apex. Although there is little information available on seed dispersal for *Scolopia chinensis*, the closely related species *Scolopia saeva*, which has very similar berries, is known to be dispersed by birds in Hong Kong.[62]

Scolopia chinensis is treated here as a member of the plant family Salicaceae. Historically, the Salicaceae was restricted to the poplars (*Populus*) and willows (*Salix*), which are characterised by their wind-pollinated catkins composed of highly reduced, unisexual flowers. Recent molecular phylogenetic research has necessitated a fundamental reappraisal of the taxonomy, however, since the *Populus-Salix* lineage has been shown to be nested within part of the Flacourtiaceae family.[221] The name Salicaceae has been adopted for this revised family circumscription, and all Hong Kong representatives of the 'Flacourtiaceae' (including *Homalium cochinchinense, q.v.*) are now transferred to the significantly enlarged Salicaceae.[12]

Plate 94. *Scolopia chinensis.* (A) Flowering branch, showing flowers with elongate stamens. (B) Single flower, showing tetramerous sepals and petals. (C) Fruiting branch with red berries and their persistent stigmas. (D) Branch, with stout spines. (E) Leaf base, showing paired glands on opposite side of the petiole.

A

B

C

D

E

Sinosideroxylon wightianum (Hook. & Arn.) Aubrév.

(= *Sideroxylon wightianum* Hook. & Arn.)

———— ℰ℈ ————

Sinosideroxylon wightianum (Iron olive or Wight's Sinosideroxylon; 革葉鐵欖 or 鐵欖; Sapotaceae)[343] is a small tree species (growing to *c.* 8 m) that is locally common in shrubland, secondary forest and *fung shui* woods.[1] As with most species in the family Sapotaceae, the branches and twigs exude a white latex when damaged,[344] providing protection against microbial infection. The latex in some Sapotaceae species is of economic importance: the Southeast Asian species *Palaquium gutta* is the source of 'gutta-percha,' a non-elastic polymer that becomes plastic on heating but which retains its shape after cooling;[345] and the Neotropical species *Manilkara zapota* was formerly the source of 'chicle,' which was used as the elastic component in early chewing gums.[346]

Sinosideroxylon wightianum flowers are either solitary or borne in clusters of 2–5. Each flower is generally 'pentamerous,' with five (rarely six) small sepals (*c.* 2.5 mm long) and an equal number of fused greenish-white petals (6–8 mm long) that are basally fused to form a corolla tube.[343] The apex of each petal lobe is inwardly curved over individual stamens, which are basally fused to the corolla tube and which protrude beyond the top of each petal. The five stamens are separated by sterile stamens ('staminodes') that are flattened and petal-like (although considerably smaller than the petals).

The fruits are single-seeded fleshy, deep purple berries (10–18 × 5–8 mm). Although there is little information available on frugivory and seed dispersal of *Sinosideroxylon wightianum*, the fleshy fruits of Sapotaceae species are typically primate dispersed, with the seeds either spat out or passing undamaged through the animal's gut.[344] A closely related species from Mauritius (classified in *Sideroxylon*, a genus that often encompasses *Sinosideroxylon*) was previously believed to have been obligately dispersed by dodos (*Raphus cucullatus*), with the seeds requiring abrasion in the dodo's gizzard before being able to germinate.[347] Although this would explain the apparent rarity of natural seedlings after 1681 when the dodo was exterminated, the theory has since been debunked: evidence now suggests that many trees are younger than 300 years old, and it is likely that the seeds are dispersed by tortoises.[348]

Plate 95. *Sinosideroxylon wightianum.* (A) Flowering branch, bearing small greenish-white flowers. (B) Single flower, with five petals that are inwardly curved to partially enclose the five protrudiung stamens. (C) Cluster of deep purple berries.

A

B

C

Sloanea sinensis (Hance) Hemsl.

(= *Echinocarpus sinensis* Hance; *Sloanea hongkongensis* Hemsl.)

―――― ᠊᠊ᢒᠯᠣ᠊ ――――

Sloanea sinensis (Chinese Sloanea; 猴歡喜 or 香港猴歡喜; Elaeocarpaceae)[187] is a large native tree species (growing to *c.* 20 m in height) that is locally common in secondary forests.[1] The flowers are borne in small axillary inflorescences, with individual flowers oriented downwards due to their often noticeably angled pedicel (flower stalk). Each flower has four ovate sepals (6–8 mm long) and four yellowish-white petals (7–9 mm long) that are irregularly lobed at the apex. The flowers are bisexual, with 60–80 stamens arising from a flattened glandular disc that encircles the base of the compound pistil.[349]

Although *Sloanea sinensis* belongs to the same plant family as *Elaeocarpus chinensis* (*q.v.*), the two species are strikingly different in their fruit morphology: whereas the latter has fleshy 'drupes', *Sloanea sinensis* forms dry, dehiscent 'capsules' that are adorned with stiff bristles, 10–15 mm long. The capsules split into 3–7 distinct segments that are internally purple at maturity, exposing up to four shiny, black seeds (10–13 mm long) per segment. The upper half of each seed is capped with a fleshy yellow 'aril' that clearly serves as an attraction and/or reward to potential seed dispersers. Although there is little information available on patterns of frugivory or seed dispersal in *Sloanea sinensis*, there is an extensive literature on other species in the genus indicating dispersal by birds[270,350] and primates.[68]

It was the discovery of a dehisced *Sloanea* capsule with its brightly coloured arillate seed in Singapore in July 1944 that led the resident botanist E. J. H. Corner (1906–96) to speculate on the evolution of these fruits. His conclusion— that it is unlikely that such complex arils would have evolved independently in multiple different plant families—led him to propose his 'Durian theory', in which he suggested that the combination of dehiscent fruits with arillate seeds is the primitive condition within flowering plants.[351] Although Corner's thoery has not withstood the rigours of subsequent scientific investigation and is not supported by contemporary phylogenetic reconstructions, it nevertheless proved to be profoundly influential, foreshadowing contemporary views on the importance of plant-animal coevolution.

―――――――――――――――――――――――――――――――――――――

Plate 96. *Sloanea chinensis.* (A) Flowering branch and isolated flower. (B) Ripe fruit with its stiff bristles. (C) Open fruit capsule, showing purple interior. (D) Shiny, black seeds.

Sterculia lanceolata Cav.

———— ❧ ————

Sterculia lanceolata (Lance-leaved Sterculia or Scarlet Sterculia; 假蘋婆 or七姐果; Malvaceae)[213] is a common species in lowland forests, especially near streams. As with many local trees, *Sterculia lanceolata* shows seasonality in leaf growth, with old leaves shed and replaced by new growth each year; unlike most species, however, *Sterculia lanceolata* undergoes this 'leaf exchange' very late in the year, and is generally leafless in mid-summer when most other species bear young leaves.[2] It has been suggested that this may be because in Hong Kong the species occurs close to its northern latitudinal extreme, and that leaf formation and reproduction are restricted to the months that are hottest and wettest, and most similar to the tropical conditions experienced by the species elsewhere.[2]

The flowers are small (*c.* 1 cm in diameter) but are borne in conspicuous inflorescences. The sepals are basally fused to form a tube and are pinkish-red and petal-like, although true petals are absent. The flowers are unisexual with both sexes borne on the same tree. The stamens in the male flowers are fused to form a column with the anthers (which contain the pollen) aggregated in a terminal spherical cluster (the 'androgynophore').[213] The female flowers possess five carpels surrounded by a ring of sterile stamens and hence appear to be structurally bisexual although they are functionally female.

The fruits consist of five bright scarlet and leathery 'follicles' that develop from the essentially unfused carpels. The follicles split open at maturity to reveal prominent shiny black seeds, which are eaten and dispersed by birds.[62] Unlike most animal-dispersed seeds, it is the seeds of *Sterculia lanceolata* rather than the fruit itself that constitute the food reward for the animal. The seed coat has three distinctive layers: a thin outer black pigmented layer; a thin, white pulpy middle layer, which is presumably of nutritive value to the birds despite being so small; and a tough inner layer that protects the embryo inside the seed.[68]

Plate 97. *Sterculia lanceolata.* (A) Flowering branch. (B) Flowers with petal-like sepals. (C) Dissected flowers. (D, E) Fruits (green when immature, opening and turning scarlet at maturity), revealing the prominent shiny black seeds.

Styrax suberifolius Hook. & Arn.

—— ℰℛ ——

Styrax suberifolius (Cork-leaved snow-bell; 栓葉安息香 or 紅皮; Styracaceae)[352] is an attractive ornamental tree with pendent clusters of flowers. Each flower is bisexual, with a cup-shaped calyx formed by the fusion of five sepals, and a corolla comprising five white petals (up to 12 mm long) that are basally fused. The flowers have between eight and ten stamens (C in Plate 98) with conspicuous yellow pollen-bearing anthers, and a pistil derived from the fusion of three carpels (D in Plate 98). The flowers are visited by *Apis cerana* and *Xylocopa* bees,[61] which are effective pollinators.

The leaves are very distinctive, with a striking colour contrast between the green upper surface and brown lower surface. The underside of the leaf is densely covered with branched brown hairs, and superficially appears 'cork-like', hence the specific epithet (which literally translates as 'cork-leaved') and the English vernacular name for the tree.

Styrax suberifolius is locally common in secondary forests in Hong Kong, in contrast with other local representatives in the genus (*Styrax confusus* and *Styrax odoratissimus*), which are both much rarer.[1] This difference might be explained by contrasting fruit structure: all *Styrax* species have large, dry fruits that lack any outer fleshy layer, but amongst the three Hong Kong species only *Styrax suberifolius* has fruits that open to release the seeds.[352] The two rarer species are likely to be dispersed by scatter-hoarding rodents,[2] which collect dry fruits and seeds and cache them for later consumption; subsequent failure to retrieve seeds would enable successful germination. The scatter-hoarding rodents that naturally disperse such seeds are now locally extinct,[2] however, and as a result trees that rely on this dispersal mechanism are inevitably rare in younger secondary forests, generally only persisting in older, more established forests. It can be hypothesised that *Styrax suberifolius* has escaped this dispersal constraint by releasing its seeds from the fruit, enabling an alternative dispersal mechanism (although exising knowledge is limited).

Plate 98. *Styrax suberifolius.* (A) Flowering branch, showing flowers with conspicuous yellow anthers. (B) Flower. (C) Stamens. (D) Carpel. (E) Fruiting branch. (F) Open fruit.

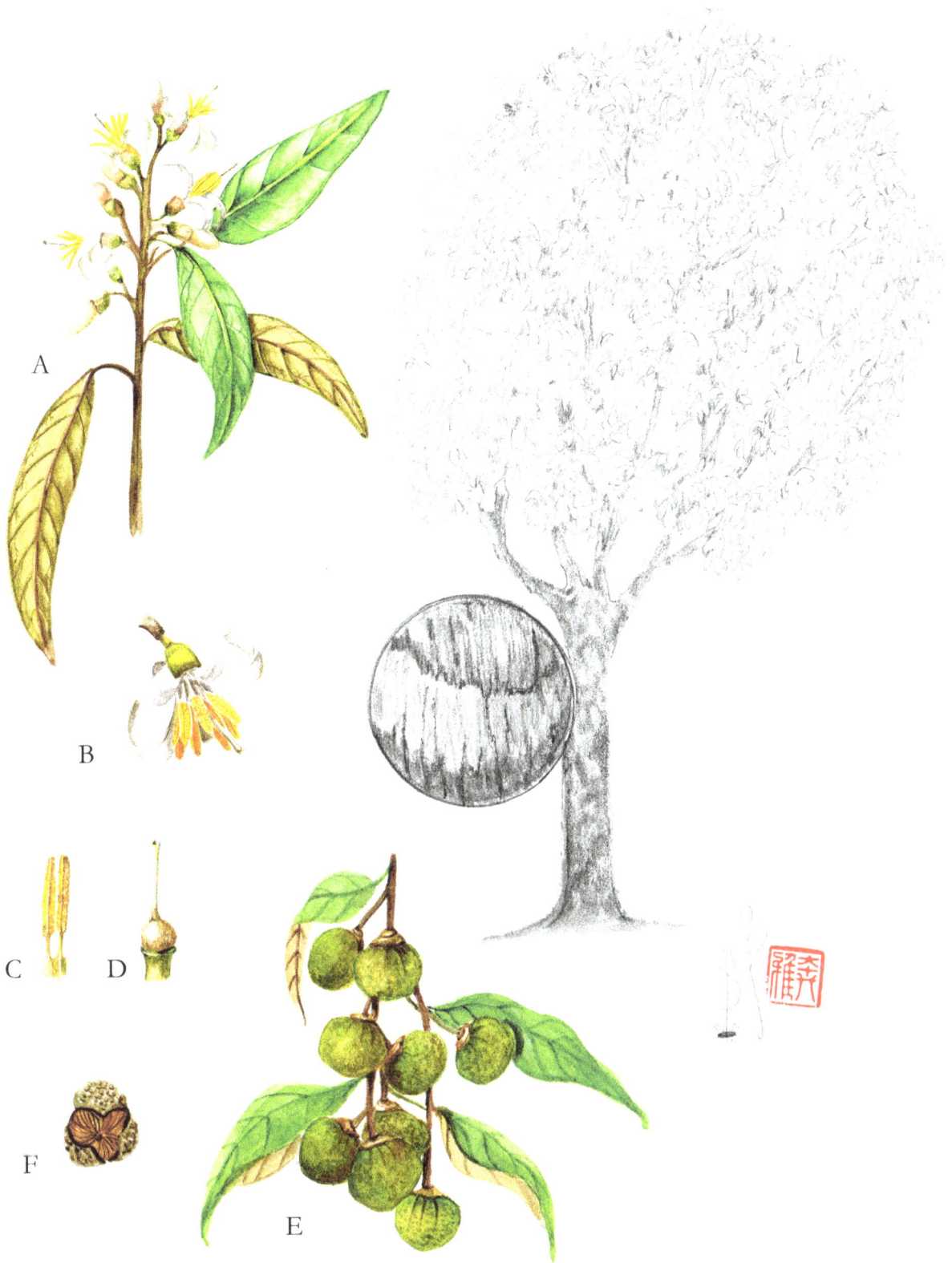

Symplocos congesta Benth.

———— ❦ ————

Symplocos congesta (Dense-flowered sweet-leaf or Congested Symplocos; 密花山礬; Symplocaceae)[353] belongs to a comparatively large genus with *c.* 300 species that are widely distributed in tropical and subtropical forests. Fourteen *Symplocos* species are recorded from Hong Kong: *Symplocos congesta* can easily be differentiated from the others as the flowers have a very short stalk and are aggregated into tight clusters (hence explaining the scientific epithet '*congesta*' and the English vernacular names).

Symplocos congesta flowers are small and bisexual, with five white petals that are basally united, up to *c.* 50 stamens that are fused into clusters that are themselves fused to the base of the petals, and a compound pistil derived from the fusion of three carpels.[353] Although the pollination ecology of *Symplocos congesta* is unknown, other local species are reported to be largely bee-pollinated but also to be visited by other insect groups, including flies, butterflies, moths and beetles.[61] The lack of knowledge of the reproductive biology of the genus is surprising given its cosmopolitan distribution in tropical and subtropical regions. Many South American species have been shown to exhibit 'cryptic dioecy,' in which individuals bear morphologically bisexual flowers that are functionally either male (with sterile carpels) or female (with sterile stamens), thereby preventing self-fertilisation.[354] A similar mechanism has recently been reported in a species in Taiwan,[355] but so far no studies have been conducted on the Hong Kong species.

The fruits are small, purplish-blue 'drupes'—fleshy berry-like fruits with a hard inner fruit wall surrounding the seeds. Seed dispersal has not been observed in *Symplocos congesta*, although it has been inferred that the fruits are likely to be consumed by birds:[62] the flesh of the fruit is glucose-rich as is typical for bird-dispersed species.[63]

Plate 99. *Symplocos congesta*. (A) Flowering branch, showing dense clusters of small, white flowers. (B, C) Purplish-blue fruits.

Syzygium jambos (L.) Alston
(= *Eugenia jambos* L.)

———— ℘ ————

Syzygium jambos (Rose apple; 蒲桃; Myrtaceae)[254] is native to tropical Southeast Asia but was introduced to Hong Kong in 1860,[241] presumably because of its edible fruits.[356] It is commonly found growing in *fung shui* woods close to villages and is now fully naturalised in Hong Kong, with well-established self-sustaining populations.[357]

The flowers are 3–4 cm in diameter, with conspicuous cream-coloured petals up to 1.5 cm long.[254] The sepals and petals are basally fused, forming a structure known as the 'hypanthium' that surrounds the fused ovary (the base of the carpel) and hence protects the ovules. Each flower is bisexual, with numerous long stamens (2–2.8 cm) and a style of similar length. The flowers are pollinated by a variety of birds and bats, which feed on the nectar.[51]

The fruits of *Syzygium jambos* are large (up to 5 cm in diameter) and fleshy, with a greenish-yellow skin that is sometimes flushed pink. The fruits are eaten by bats, which have been observed to carry the fruits to their feeding roosts up to 40 m from the maternal tree for processing prior to consumption.[51] The seeds are large (*c.* 18 mm in diameter) and are rejected by the bats and dropped immediately below the feeding roosts; it therefore appears that *Syzygium jambos* is primarily dispersed by bats in Hong Kong.[51] The trees are often restricted to areas close to rivers, although it is unclear whether this is due to the bats' flight patterns or factors that limit seedling establishment.[110]

Plate 100. *Syzygium jambos.* (A) Flowering branch. (B) Solitary flower, showing calyx. (C) Fruits.

A

B

C

Tetradium glabrifolium (Champ. ex Benth.)
T. G. Hartley
(= *Euodia meliifolia* (Hance ex Walp.) Benth., as '*meliaefolia*')

—— ∾ ——

Tetradium glabrifolium (Melia-leaved Evodia; 棟葉吳茱萸; Rutaceae)[60] is locally common along forest margins and in disturbed sites. The species was previously classified in the genus *Euodia* (often erroneously spelt *Evodia*), together with another local species that has subsequently been renamed *Melicope pteleifolia*. These two species are similar but can easily be distinguished based on leaf shape (both have divided leaves, but *Melicope pteleifolia* has a maximum of three leaflets per leaf, whereas *Tetradium glabrifolium* has up to 11) and floral sex (bisexual in *Melicope pteleifolia*, unisexual in *Tetradium glabrifolium*).[60,358]

Tetradium glabrifolium flowers are very small and borne in large inflorescences, with inconspicuous white petals (*c.* 3 mm long). Each flower has organs arranged in whorls of either four or five: although the generic name *Tetradium* suggests a 'tetramerous' arrangement (in whorls of four), the floral organs are more commonly arranged in whorls of five ('pentamerous'). The flowers are functionally unisexual in order to avoid the possibility of within-flower self-fertilisation; the evolutionary origin of this unisexuality from ancestors that had bisexual flowers is clearly indicated by the vestigial sterile carpels (known as 'pistillodes') that occur in male flowers, and the corresponding sterile stamens ('staminodes') in female flowers.

The flowers are primarily pollinated by *Apis cerana* bees, but are also visited by other insect groups, including wasps, butterflies and beetles.[61] After fertilisation, the individual carpels in the flower develop into small dry single-seeded 'follicles' (*c.* 5 mm in diameter) that split open along the upper margin, exposing the seed (Plate 102). Although each carpel in the flower has two ovules, only one of these develops into a seed in the follicle.[358] In *Tetradium glabrifolium*, the seeds remain attached to the follicle walls and are dispersed by birds;[157,350] this contrasts with the closely related local species *Melicope pteleifolia*, in which the seeds are ejected entirely from the fruits.[68]

Plate 101. *Tetradium glabrifolium* (fruiting branch shown in Plate 102). (A) Flowering branch, showing inflorescences composed of small, white flowers. (B) Single male flower, with fertile stamens and sterile pistillode.

A

B

Plate 102. *Tetradium glabrifolium* (flowering branch shown in Plate 101). (A) Fruiting branch, with clusters of fruits. (B) Unopen follicles. (C) Open follicles, with single exposed seed in each follicle.

Thespesia populnea (L.) Sol. ex Corr.

——— ✍ ———

Thespesia populnea (Portia tree; 繖楊 or 恒春黃槿; Malvaceae)[215] is widespread in coastal areas in the tropics, although it is relatively uncommon in Hong Kong. The naturalist Daniel Solander first collected the species from the South Pacific island of Tahiti during Captain Cook's circumnavigation of the globe in *H. M. S. Endeavour* (1768–71). Solander apparently coined the generic name 'Thespesia'—derived from the Greek word for 'divine'—because the Tahitians regarded the trees as sacred.[74]

Thespesia populnea grows well in sandy soils and is able to withstand highly saline conditions resulting from sea spray. Although it is therefore ideal for cultivation as a shade tree behind beaches, *Thespesia populnea* has not been widely planted in Hong Kong.[80]

The attractive yellow flowers of *Thespesia populnea* resemble those of *Hibiscus tiliaceus* (*q.v.*), with stamens that are fused into a cylinder that surrounds the united styles of the fused carpels. The flowers are comparatively short-lived, turning purple-red as they age due to the accumulation of anthocyanin pigments,[359] although the impact of this colour change on pollinator attraction is unknown.

The fruits are essentially indehiscent (in marked contrast with *Hibiscus tiliaceus*) and are dispersed by sea: after several weeks afloat, the fruit capsule disintegrates to release the seeds.[68,155] The seeds are furthermore buoyant because of air gaps between the immature seed leaves ('cotyledons'), and are known to be able to germinate after immersion in seawater for up to a year and even after exposure to sub-zero temperatures.[68] *Thespesia populnea* is therefore well adapted to long-distance dispersal in which ocean currents transport the seeds through cold southern waters; this presumably explains the widespread distribution of the species around the Indian and Pacific Oceans.

Plate 103. *Thespesia populnea.* (A) Flowering branch, showing a younger yellow flower alongside an older, purple-red flower. (B) Mature indehiscent fruit capsule. (C) Older fruit that has begun to disintegrate, releasing the seeds. (D) Seeds.

Toxicodendron succedaneum (L.) O. Kuntze

(= Rhus succedanea L.)

——— ✥ ———

Toxicodendron succedaneum (Japanese lacquer tree or Wax tree; 木蠟樹 or 野漆樹; Anacardiaceae)[329] is a small tree, generally growing to 5 m (rarely to 12 m), that is common in shrubland and early secondary forests in Hong Kong.[1] The species is widely known in the local botanical literature, including the *Flora of Hong Kong*,[329] under the synonym *Rhus succedanea*. Recent phylogenetic analyses of DNA sequence data, however, has revealed that *Rhus* and *Toxicodendron* form two separate evolutionary lineages and hence are better treated as distinct genera;[360] since the species illustrated here is deeply nested within the latter lineage it is clear that the name *Toxicodendron succedaneum* should be used.

The generic name *Toxicodendron* literally means 'poisonous tree', reflecting the toxicity of all species in the genus, including *Toxicodendron succedaneum*[361] and the notorious Poison ivy, *Toxicodendron radicans*.[362] *Toxicodendron* species contain a chemical known as 'urushiol' that can cause serious skin irritation, with the formation of large blisters; interestingly, people who are sensitive to *Toxicodendron* sap can develop a rash after handling mangoes (*Mangifera indica*),[363] which belong to the same plant family.

A commercially important resin that contains urushiol is obtained from the sap of *Toxicodendron succedaneum* (and the closely related species, *Toxicodendron vernicifluum*). After prolonged oxidation and polymerisation, the resin forms a hard and durable lacquer that has been used in the manufacture of artifacts in China and Japan for many centuries.[364]

The flowers of *Toxicodendron succedaneum* are small (*c.* 2 mm in diameter), yellowish-green and borne in large inflorescences, and are commonly visited by *Apis cerana* bees.[61] The fruits are small 'drupes' (with a tough inner fruit wall, surrounding the seed) that are eaten and dispersed by birds.[365] Although the drupes are rather dull coloured, their maturation coincides with leaf senescence towards the end of the year, and it seems likely that the birds are attracted to fruit-bearing trees by the bright red colouration of senescent leaves.[2,62]

Plate 104. *Toxicodendron succedaneum.* (A) Flowering branch, with large inflorescence composed of small, yellowish-green flowers. (B) Fruiting branch, showing bright red senescent leaves that coincide with fruit maturation. (C) Immature fruits.

A

B

C

Trema tomentosa (Roxb.) Hara

(= *Sponia velutina* Planch.; *Trema amboinensis* (Willd.) Blume;
Trema orientalis auct. non (L.) Blume)

———— ✃ ————

Trema tomentosa (India-charcoal Trema; 山黃麻; Cannabaceae)[150] is a medium-sized tree (growing to *c.* 10 m) that is widespread across Southeast Asia and locally common in shrubland and disturbed habitats in Hong Kong. It is a pioneer species with seedlings that are very shade intolerant;[316] the species nevertheless has potential for afforestation in the region as it grows rapidly, enabling the early formation of a forest canopy, and quickly maturing to produce fruits that are attractive to birds.[316]

Many parts of the tree are densely hairy (or 'tomentose'), including the young branches and leaves; this hairiness is reflected in the specific epithet '*tomentosa*.' The flowers are small and unisexual, with five sepals but lacking petals, and are aggregated into inflorescences (2–4.5 cm long) that comprise flowers of only one sex.[150] The functionally male flowers contain five stamens and a rudimentary carpel that is sterile, whereas the functionally female flowers have a fertile carpel but lack any vestigial stamens. Pollination ecology studies of *Trema tomentosa* populations in Borneo have shown that the flowers are frequently visited by stingless bees belonging to the genus *Trigona*.[337]

Trema tomontosa was traditionally classified in the elm family, Ulmaceae. Recent molecular phylogenetic research—in which the evolutionary tree of the group is reconstructed by comparing the species' DNA sequence—has revealed, however, that true elm trees (in the genus *Ulmus* and their close relatives, which have bisexual flowers and dry fruits) form a separate evolutionary lineage from the genus *Trema* and its relatives (with unisexual flowers and fleshy fruits).[366] The latter lineage has accordingly been transferred to the family Cannabaceae,[12] which was formerly restricted to species of *Cannabis* (the source of the psychoactive drug of the same name, as well as hemp fibres used for making ropes) and *Humulus* (the source of hops used for brewing beer).

Plate 105. *Trema tomentosa.* (A) Flowering branch, showing inflorescences composed of small flowers that lack petals. (B) Lower surface of leaf. (C) Flowers. (D) Fruiting branch.

A

D

B

C

Turpinia montana (Blume) Kurz

(= *Turpinia cochinchinensis* auct. non (Lour.) Merr.;
Turpinia nepalensis auct. non Wall. ex Wight & Arn.;
Turpinia pomifera auct. non DC.)

——— ☙ ———

Turpinia montana (Turpinia; 山香圓; Staphyleaceae)[367] is a shrub or small understory tree (growing to *c.* 12 m) that is locally common in secondary forests in Hong Kong.[1] It has compound leaves comprising three or five (rarely seven) opposing leaflets with a terminal leaflet; each leaflet has a serrated margin, with conspicuous forward-pointing teeth.

The small flowers are loosely borne in 8–12 cm-long inflorescences. Each flower is 'pentamerous' (with organs in whorls of five), with five tiny sepals (*c.* 1.3 mm long), five yellowish petals (*c.* 2 mm long), five stamens alternating with the petals, and a three-chambered compound pistil.[367] The fruits are small rounded berries (4–7 mm in diameter), each containing three or four seeds.

Evolutionary relationships within the family Staphyleaceae have recently been reassessed using DNA sequence data.[368] This has revealed several major surprises: although the three genera traditionally included in the family (*Euscaphis*, *Staphylea* and *Turpinia*) are easily delimited using morphological characters, the molecular data indicates that *Staphylea* and *Turpinia* are both represented by more than one evolutionary lineage. This implies that significant evolutionary convergence is likely to have occurred in the morphological characters previously used to delimit the genera: recent classifications of the family have accordingly proposed the recognition of two newly circumscribed genera, *Dalrympelea* and *Staphylea*,[369] with *Turpinia montana* presumably to be included in the former.

The other surprise arising from the molecular phylogenetic analysis of the Staphyleaceae is the suggestion that the New World lineage within *Turpinia* may have arisen as a result of ancient hybridisation between ancestors of *Euscaphis* and *Turpinia montana*.[368] If this inference is correct, the hybridisation event would likely have occurred in China (where *Euscaphis* and *Turpinia montana* co-exist), with subsequent dispersal of the hybrid lineage to the New World.

Plate 106. *Turpinia montana.* (A) Flowering branch, with an inflorescence of small flowers. (B) Single flower, showing its five yellowish petals, five stamens and compound pistil. (C) Fruiting branch, with small berries.

A

B

C

Vernicia montana Lour.

(= *Aleurites montana* (Lour.) E. H. Wilson;
Aleurites cordata auct. non (Thunb.) R. Br. ex Steud.)

———— ❧ ————

Vernicia montana (Wood-oil tree; 木油桐; Euphorbiaceae)[77] and its close relative, *Vernicia fordii* (Tung-oil tree; 油桐 or 三年桐), are widely cultivated for oils that are extracted from the seeds for use as a wood varnish and for burning in traditional oil lamps.[370] Both species were incorporated in the Hong Kong Government's pre-war reafforestation scheme, but with only limited success.[25] Although these species are currently only of limited commercial importance there is nevertheless considerable interest in their development as a source of biodiesel fuel.[371]

Vernicia montana is closely related to *Aleurites moluccana* (*q.v.*) from which it differs in having much larger flowers and lacking star-shaped hairs on the young branches.[372] As with most species in the Euphorbiaceae family, the leaves possess a pair of conspicuous extra-floral nectaries at the apex of the petiole;[49] these glands provide sweet nectar for ants, which in turn provide some protection against herbivory. The two *Vernicia* species cultivated in Hong Kong can be distinguished by the shape of these glands, which are stalked in *Vernicia montana* but sessile in *Vernicia fordii*.[77]

The flowers develop in large, showy inflorescences. Each flower is unisexual, with both sexes of flower either borne on the same tree ('monoecy') or with the entire tree bearing flowers of only one sex ('dioecy'). The flowers have five white petals (2–3 cm long) with purple-red stripes towards the base, and small glands that alternate with the petals. The male flowers have 8–10 stamens in two whorls (the inner of which are basally fused), whereas the female flowers have a pistil with three bilobed stigmas.

Plate 107. *Vernicia montana.* (A) Flowering branch, with large inflorescences of white flowers. (B) Single male flower. (C) Pistil with three bilobed stigmas. (D) Paired glands at the apex of the petiole. (E) Stamen. (F, G) Immature and mature fruits.

A

B

C

D

E

F

G

Viburnum odoratissimum Ker Gawl.

——— ❧ ———

Viburnum odoratissimum (Sweet Viburnum; 珊瑚樹; Adoxaceae)[373] is a small tree, growing to 10 m (rarely to 15 m), common in young secondary forests in Hong Kong.[1] The trees bear large inflorescences (up to 14 cm long) of small, fragrant flowers. Each flower has five whitish to yellowish petals that are fused into a short tube, *c.* 2 mm long. The flowers are visited by a broad range of insects—including bees, wasps, calliphorid flies, pierid butterflies and beetles[61]—which are attracted by the sweet floral scent.

The fruits are small 'drupes' (fleshy berry-like fruits with a hard inner fruit wall, surrounding the solitary seed). These fruits undergo a significant colour change as they mature, changing from green when immature to red (B in Plate 108) and then blue-black when fully mature (C in Plate 108). It is not surprising that the fruits are known to be eaten by birds[62] as these features are all characteristic of avian frugivory and seed dispersal.

Viburnum odoratissimum appears to be relatively flame resistant and is therefore commonly planted locally as a fire-break to help contain hill fires.[325,374] It appears that the seeds have evolved the capacity to remain dormant while environmental conditions are unsuitable for seed germination. This dormancy has been shown to be 'morphophysiological,' in which the development of the embryo is retarded due to the presence of chemical inhibitors;[375] it is only once these chemicals are broken down in older seeds that the embryo can develop and become strong enough to break through the tough seed coat.

The leaves of other *Viburnum* species are known to contain toxic compounds that inhibit or suppress the growth of neighbouring plants;[376] this phenomenon—known as 'allelopathy'—is observed in many plants and is believed to have evolved to reduce competition for nutrients. Fishermen in Japan have traditionally used the toxic leaves of *Viburnum odoratissimum* as a 'fish-stupefying agent' to poison fish: the leaves were macerated and introduced into tidal pools that had been enclosed to prevent the fish from escaping.[377,378]

Plate 108. *Viburnum odoratissimum.* (A) Flowering branch, with a large inflorescence composed of small whitish flowers. (B) Isolated flowers. (C) Fruiting branch with red fruits, soon after maturation. (D) Fruiting branch with fully mature, blue-black fruits.

Zanthoxylum avicennae (Lam.) DC.

(= *Zanthoxylum lentiscifolium* Champ.)

——— ❧ ———

Zanthoxylum avicennae (Prickly ash; 簕欓花椒 or 簕欓; Rutaceae)[60] is a common local tree, growing to *c.* 15 m. The leaves are compound, with 13–18 (sometimes up to 25) leaflets, and are reminiscent of those of Ash trees (*Fraxinus* species, in the distantly related family Oleaceae[379]); although the English vernacular name reflects this similarity it also highlights the profusely spiny branches that clearly distinguish it from a true Ash.

Zanthoxylum avicennae flowers are small and borne in large inflorescences. Each flower consists of five green sepals and five pale yellow petals. The flowers are unisexual: male flowers have five fertile stamens and a sterile carpel (known as a 'pistillode'); whereas the female flowers have two or three fertile carpels and small sterile stamens ('staminodes').

After fertilisation, the carpels develop into small, highly aromatic fruits that turn purple-red at maturity. Although *Zanthoxylum avicennae* fruits are not cultivated for culinary purposes, other species in the genus are the source of 'Sichuan pepper', which has a pungent peppery taste causing the slight 'numbing' sensation characteristic of Sichuanese cuisine. *Zanthoxylum avicennae* fruits are now the focus of pharmacological research aimed at evaluating its use as an anti-inflammatory drug.[380]

Zanthoxylum avicennae fruits split open at maturity to expose the glossy black seeds, which remain suspended from the fruit wall. The fruits are eaten by birds, which are effective dispersers of the seeds.[103]

Plate 109. *Zanthoxylum avicennae.* (A) Flowering branch, with inflorescences of small flowers. (B) Male flower. (C) Female flower. (D) Infructescence of small fruits. (E) Mature fruit, split open to expose its seed.

Glossary

———— ✂ ————

Abaxial. The surface facing away from the vertical axis during development (hence in leaves, the lower surface).

Acorn. A dry fruit consisting of a nut (*q.v.*) that is nested within a basal cupule (*q.v.*) (observed, for example, in *Lithocarpus glaber*).

Adaxial. The surface facing towards the vertical axis during development (hence in leaves, the upper surface).

Allelopathy. The chemical inhibition of one species by another (observed, for example, in *Acacia confusa*).

Androdioecy (adj. androdioecious). With separate male (staminate) and bisexual flowers that are borne on different individuals.

Androecium. Collective term for the stamens in a flower.

Androgynophore. An elongated extension of the receptacle (*q.v.*) in the flower, bearing the stamens and carpels (observed, for example, in *Reevesia thyrsoidea*).

Angiosperms. Flowering plants.

Anther. The upper, pollen-bearing part of a stamen, generally borne on an elongated filament.

Aphids. Small sap-sucking insects belonging to the family Aphididae.

Aril. A fleshy extension of the outer layer of the seed coat, functioning as a food reward for seed dispersers (observed, for example, in *Acacia auriculiformis*).

Auct. non. An abbreviation for the Latin phrase 'auctores non' (= 'authors other than'): commonly used after the citation of species names, indicating that the name has been applied in ways not intended by the taxonomist who first coined the name.

Axil. The angle between a stem and a leaf (or other organ).

Axillary. With reference to the position of an organ, arising in an axil (*q.v.*).

Berry. An indehiscent fleshy fruit with one to many seeds; distinguished from 'drupes' (*q.v.*) by the absence of a tough inner fruit wall.

Binomial. A scientific name used for species, composed of two parts: the generic name followed by the specific epithet (*q.v.*).

Bipinnate. Consisting of leaflets (pinnae) that are themselves divided into higher-order leaflets.

Bisporangiate. (With reference to stamens) having two pollen-bearing chambers.

Bract. A modified (often highly reduced) leaf-like structure.

Calliphorid flies. Flies such as blue-bottles, belonging to the family Calliphoridae.

Calyx. Collective term for the sepals in a flower.

Capsule. A dry fruit that develops from several fused carpels, and splits open at maturity to release its seeds.

Carpel. The female reproductive organ in a flower (comprising stigma, style and ovary), sometimes fused together to form a pistil (*q.v.*).

Caruncle. A type of aril (*q.v.*): a fleshy outgrowth of the seed coat associated with animal dispersal of seeds (observed, for example, in *Aquilaria sinensis*).

Chlorophyll. The green pigment in plant cells that enables photosynthesis.

Compound leaf. A leaf that is subdivided into smaller units (leaflets or pinnae).

Corolla. Collective term for the petals in a flower.

Cotyledons. The first-formed leaves in a seedling (seed leaves, generally morphologically distinct from the mature leaves).

Cultivar. A variety of plant produced and maintained in cultivation (*i.e.*, not the result of natural evolution).

Cupule. A cup-shaped involucre (*q.v.*) formed from fused bracts (observed, for example, in *Castanopsis fissa*).

Dehiscence. The splitting open of a structure at maturity (*e.g.*, anthers that open to release pollen grains, and fruits that open to release seeds).

Dicotyledons. Flowering plants characterised by the possession of a two seed leaves (cotyledons); often abbreviated as 'dicots.'

Dioecy (adj. dioecious). With separate male (staminate) and female (pistillate) flowers that are borne on different individuals.

DNA. Deoxyribonucleic acid: complex molecules responsible for carrying genetic information in cells.

Drupe. An indehiscent fleshy fruit with a hard inner layer to the fruit wall.

Endemic. Restricted to a specific geographical location.

Epimatium (pl. epimatia). A swollen appendage derived from the ovuliferous bract in the female reproductive structures of *Podocarpus* (observed, for example, in *Podocarpus macrophyllus*).

Exotic. Describes a species (or other taxonomic group) that is not native to a specified geographical region (*i.e.*, introduced as a result of human activity).

Extra-floral. Located outside the flower (*e.g.*, extra-floral nectaries, observed on the leaves of many species in the family Euphorbiaceae).

Family. A category in biological classifications, consisting of related genera.

Fascicle. A cluster of similar organs.

Fasciclode. A cluster of sterile stamens (observed, for example, in *Cratoxylum cochinchinense*).

Fig wasps. Small wasps associated with figs, belonging to the family Agaonidae.

Filament. The stalk of a stamen, supporting the anther.

Flower bugs. Hemipteran insects belonging to the family Anthocoridae.

Follicle. A dry, dehiscent fruit that develops from a single, unfused carpel.

Frugivore. An animal that eats fruits (and hence contributes to seed dispersal).

Fung shui (= ***feng shui***) **woods.** Traditionally protected woodlands maintained in close proximity to villages in accordance with the Chinese system of geomancy.

Gall. A plant growth in response to a parasite (e.g., bacterial, fungal and insect); galls often contain an insect larva, and the shape of the gall can be characteristic of both the plant and insect species involved.

Gall midges. Small flies belonging to the family Cecidomyiidae, responsible for the formation of many plant galls.

Gamete. The mature reproductive cell (male sperm or female egg).

Genus (pl. genera). A category in biological classifications, consisting of related species.

Gymnosperms. Cone-bearing plants, such as pine trees (genus *Pinus*).

Gynoecium. Collective term for the carpels in a flower.

Hermaphroditic. Used to describe flowers that are bisexual, bearing both stamens and carpels.

Hesperiid butterflies. Butterflies, including Skippers, belonging to the family Hesperiidae.

Homologous. Used to describe structures that share a common ancestry, although not necessarily retaining similar functions or appearance.

Hypanthium. A cup-shaped extension of the receptacle (*q.v.*) of the flower, generally also resulting from the fusion of other floral organs, including the sepals and petals (observed, for example, in *Syzygium jambos*).

Hypocotyl. The lower part of the shoot of the seedling, located below the seed leaves (cotyledons, *q.v.*).

Indigenous. Describes a species (or other taxonomic group) that is native to a specified geographical region (*i.e.*, not introduced as a result of human activity).

Inferior ovary. Floral ovaries that are located below the point of attachment of the perianth.

Inflorescence. A cluster of flowers developing together as a single structure.

Infructescence. A cluster of fruits (derived from an inflorescence), developing together as a single structure (observed, for example, in *Liquidambar formosana*).

Involucre. A whorl or bracts at the base of a flower or inflorescence (observed, for example, in *Rhodoleia championii*).

Littoral. Associated with the seashore or margins of a lake.

Lycaenid butterflies. Butterflies, including Blues, Coppers and Hairstreaks, belonging to the family Lycaenidae.

Monocotyledons. Flowering plants characterised by the possession of a single seed leaf (cotyledon); often abbreviated as 'monocots.'

Monoecy (adj. monoecious). With separate male (staminate) and female (pistillate) flowers that are borne on the same individual.

Mutualism. A symbiosis (*q.v.*) in which two dissimilar organisms co-exist to the benefit of both species (observed, for example in the relationship between *Ficus* species and fig wasps).

Naturalised. A species that is not native to a region, but which has become successfully established.

Nectary. Glandular organ responsible for secreting nectar, either within flowers (floral nectary) or outside flowers (extra-floral nectary).

Neotropical. Belonging to the New World tropics (*i.e.*, Central and South America).

Nitidulid beetles. Small 'sap beetles,' belonging to the family Nitidulidae.

Node. The region of a stem where leaves or other branches are attached.

Nut. A single-seeded fruit with a tough, dry shell.

Nymphalid butterflies. 'Four-footed' butterflies, belonging to the family Nymphalidae.

Ostiole. A narrow opening or aperture of a structure (observed, for example, at the apex of *Ficus variolosa* inflorescences).

Ovary. The expanded basal part of a carpel or pistil, containing the ovules.

Ovule. A complex structure in seed plants containing the egg cell.

Ovuliferous scale. Part of the female cone of a pine tree (observed, for example, in *Pinus elliottii* and *Pinus massoniana*), bearing ovules.

Panicle. A type of branched inflorescence.

Papilionid butterflies. Butterflies such as Swallowtails, belonging to the family Papilionidae.

Peltate. The attachment of a stalk to the lower surface of a flat structure (observed, for example, in petiole attachment to leaves of *Macaranga tanarius*).

Pentamerous. In groups of five (commonly used to describe the numbers of organs in each whorl within a flower).

Perianth. Collective term for the sepals and petals in a flower.

Pericarp. The fruit wall (often fleshy).

Petiole. The stalk of a leaf.

Phyllode. An expanded blade (other than a leaf lamina) that has evolved for photosynthesis (observed, for example, in *Acacia confusa*, in which it is derived from the petiole).

Phylogenetics. The science of reconstructing evolutionary trees; in contemporary research this is generally achieved using comparative DNA sequence data.

Phylogeny. An evolutionary tree, showing genealogical relationships between different lineages.

Pierid butterflies. Butterflies, including Whites and Yellows, belonging to the family Pieridae.

Pinna (pl. pinnae). Leaflet (part of a compound leaf).

Pistil. The female reproductive organ in a flower, composed of either a solitary carpel or fused carpels.

Pistillate. Pistil-bearing (hence pistillate flowers are female only).

Pistillode. Sterile (often rudimentary) carpel or pistil, often occurring in flowers that are functionally male.

Pollen grain. The male spore, containing cells that form the sperm.

Pollen sac. The part of an anther where pollen grains develop.

Pollen tube. The tube that develops from germinating pollen grains, enabling the transfer of the sperm to the ovule.

Pome. A fleshy fruit derived from a flower with an inferior ovary (*q.v.*), in which the fruit wall develops from the hypanthium (*q.v.*) (observed, for example, in *Photinia benthamiana* and commercial apples and pears).

Primary forest. Forest that has not been subject to major disturbance (*i.e.*, virgin, or old-growth forest).

Protandry. A phenomenon in which functioning of the male organs in the flower precedes that of female organs, thereby preventing self-fertilisation within the flower.

Protogyny. A phenomenon in which functioning of the female organs in the flower precedes that of male organs, thereby preventing self-fertilisation within the flower.

Rachis (pl. rachides). The central axis of a structure (*e.g.*, compound leaf or inflorescence).

Receptacle. The central region of a flower to which the various floral organs are attached.

Samara. Indehiscent, winged fruit that is adapted for dispersal by wind (observed, for example, in *Casuarina equisetifolia*).

Schizocarp. A fruit that splits into separate parts (each derived from a single carpel) at maturity (observed, for example, in *Acer sino-oblongum*).

Secondary forest. Forest that has regrown after major disturbance (often anthropogenic).

Secondary growth. The lateral thickening of stems and roots, responsible for the formation of broad tree trunks.

Semi-inferior ovary. Floral ovaries that are located partly below and partly above the point of attachment of the perianth.

Sepal. Bract-like outer sterile organ of a flower.

Sessile. Without a stalk.

Spadix (pl. spadices). A thickened inflorescence axis bearing reduced flowers, enclosed by a spathe (*q.v.*; observed, for example, in male individuals of *Pandanus tectorius*).

Spathe. A large bract enclosing a spadix (*q.v.*; observed, for example, in male individuals of *Pandanus tectorius*).

Species. The basic taxonomic unit in biological classifications. Species names are composed of two words: the generic name followed by the 'specific epithet' (*q.v.*).

Specific epithet. The second word in species names (generally an adjective).

Sporangium (pl. sporangia). A structure in which spores (including pollen grains) are produced.

Sporophyll. A fertile leaf-like structure, observed, for example, in the male pine cone (*Pinus* species).

Sphingid moths. Hawkmoths, belonging to the family Sphingidae.

Stamen. Male reproductive organ of the flower, comprising the pollen-bearing anther on a filament.

Staminate. Stamen-bearing (hence staminate flowers are male only).

Staminode. Sterile (often rudimentary) stamen, often occurring in flowers that are functionally female.

Stigma. The upper part of the carpel or pistil, which is receptive to pollen.

Stipule. Small leaf-like, spine-like or scale-like appendage (usually paired) at the base of many leaves.

Stomata. Small pores on the surface of the leaf, enabling gaseous exchange.

Style. Elongated part of the carpel or pistil, connecting the stigma and ovary.

Subspecies. A low-level taxonomic unit in biological classifications: a subdivision of a species.

Superior ovary. Floral ovaries that are located above the point of attachment of the perianth.

Syconium (pl. syconia). Specialised inflorescences of fig species, in which tiny, highly reduced flowers are borne on the inner surface of a completely invaginated chamber (observed, for example, in *Ficus variolosa*).

Symbiosis. A relationship between two dissimilar organisms that co-exist, sometimes to the benefit of only one species (parasitism) or to the benefit of both (mutualism, *q.v.*).

Syncarp. A multiple fruit, derived from the fusion of carpels from several flowers (observed, for example, in *Artocarpus hypargyreus*).

Syrphid flies. Hoverflies, belonging to the family Syrphidae.

Tepal. One of the outer sterile organs of a flower—a term used when sepals and petals are not fully differentiated (observed, for example, in *Illicium angustisepalum* and *Machilus* species).

Tetramerous. In groups of four (commonly used to describe the numbers of organs in each whorl within a flower).

Tetrasporangiate. (With reference to stamens) having four pollen-bearing chambers.

Thrips. Small winged insects belonging to the order Thysanoptera.

Tomentose. An adjective describing a dense covering of short, soft hairs.

Vivipary. The process in which seeds germinate on the maternal plant (observed, for example, in *Bruguiera gymnorhiza* and *Kandelia obovata*).

Xylem. The vascular system of plants responsible for water and nutrient transport.

—— ☙ ——

Fruiting branch of *Castanopsis fissa*

References

—— ✦ ——

1. X. Zhuang, F. Xing & R. T. Corlett (1997). The tree flora of Hong Kong: distribution and conservation status. *Memoirs of the Hong Kong Natural History Society* 21: 69–126.

2. D. Dudgeon & R. T. Corlett (1994). *Hills and Streams: An Ecology of Hong Kong.* Hong Kong University Press, Hong Kong.

3. Y. Mamiya (1983). Pathology of the pine wilt disease caused by *Bursaphelenchus xylophilus. Annual Review of Phytopathology* 21: 201–220.

4. C. Darwin (1859). *On the Origin of Species by Means of Natural Selection, or The Preservation of Favoured Races in the Struggle for Life.* John Murray, London.

5. S.-X. Yang, J.-B. Yang, L.-G. Lei, D.-Z. Li, H. Yoshino & T. Ikeda (2004). Reassessing the relationships between *Gordonia* and *Polyspora* (Theaceae) based on the combined analysis of molecular data from the nuclear, plastid and mitochondrial genomes. *Plant Systematics and Evolution* 248: 45–55.

6. S. M. Ickert-Bond & J. Wen (2006). Phylogeny and biogeography of Altingiaceae: evidence from combined analysis of five non-coding chloroplast regions. *Molecular Phylogenetics and Evolution* 39: 512–528.

7. Hong Kong Herbarium & South China Botanical Garden (2007–11). *Flora of Hong Kong.* Vols 1–4. Agriculture, Fisheries & Conservation Department, Hong Kong.

8. G. Bentham (1861). *Flora Hongkongensis.* L. Reeve, London.

9. H. F. Hance (1872). Florae Hongkongensis: a compendious supplement to Mr. Bentham's description of the plants of the island of Hong Kong. *Journal of the Linnean Society, Botany* 8: 95–144.

10. S. T. Dunn & W. J. Tutcher (1912). Flora of Kwangtung and Hong Kong (China). *Kew Bulletin of Miscellaneous Information, Additional Series* 10: 1–370.

11. The Angiosperm Phylogeny Group (1998). An ordinal classification for the families of flowering plants. *Annals of the Missouri Botanical Garden* 85: 531–553.

12. The Angiosperm Phylogeny Group (2003). An update of the Angiosperm Phylogeny Group classification for the orders and families of flowering plants: APG II. *Botanical Journal of the Linnean Society* 141: 399–436.

13. The Angiosperm Phylogeny Group (2009). An update of the Angiosperm Phylogeny Group classification for the orders and families of flowering plants: APG III. *Botanical Journal of the Linnean Society* 161: 105–121.

14. The Angiosperm Phylogeny Group (2016). An update of the Angiosperm Phylogeny Group classification for the orders and families of flowering plants: APG IV. *Botanical Journal of the Linnean Society* 181: 1–20.

15. P. Osbeck (1771). *A Voyage to China and the East Indies,* translated by J. R. Forster. Benjamin White, London.

16. J. L. Prévost (1777). *Histoire Générale des Voyages, Nouvelle Collection de Toutes les Relations de Voyages par Mer et par Terre.* Vol. 22. E. van Harrevelt & D. J. Changuion, Amsterdam.

17. R. B. Hinds & G. Bentham (1842). Remarks on the physical aspect, climate, and vegetation of Hong-Kong, China, with an enumeration of plants there collected. *London Journal of Botany* 1: 476–595.

18. B. Seemann (1857). *The Botany of the Voyage of H. M. S. Herald.* L. Reeve, London.

19. D. Dudgeon & R. T. Corlett (2004). *The Ecology and Biodiversity of Hong Kong.* Friends of the Country Parks/Joint Publishing, Hong Kong.

20. J. Hayes (1984). Hong Kong Island before 1841. *Journal of the Royal Asiatic Society, Hong Kong Branch* 24: 105–142.

21. R. T. Corlett (1997). Human impact on the flora of Hong Kong Island. *In:* N. G. Jablonski (ed.), *The Changing Face of East Asia During the Tertiary and Quaternary*, pp. 400–412. Centre of Asian Studies, The University of Hong Kong, Hong Kong.

22. S. Lockhart (1898). *Extracts from a Report by Mr. Stewart Lockhart on the Extension of the Colony of Hong Kong.* Paper laid before the Legislative Council of Hong Kong, 1899.

23. P. C.-C. Lai & K.-L. Yip (2008). Vegetation of Hong Kong: the past, present and future. *In:* Hong Kong Herbarium & South China Botanical Garden (eds), *Flora of Hong Kong.* Vol. 2, pp. xvi–xxiv. Agriculture, Fisheries & Conservation Department, Hong Kong.

24. R. D. Hill (2011). Environmental change in Hong Kong—the last 60 years. *Memoirs of the Hong Kong Natural History Society* 27: 5–62.

25. R. T. Corlett (1999). Environmental forestry in Hong Kong: 1871–1997. *Forest Ecology and Management* 116: 93–105.

26. P. A. Daley (1965). *Forestry and its Place in Natural Resource Conservation in Hong Kong: A Recommendation for Revised Policy.* Agriculture and Fisheries Department, Hong Kong.

27. L. M. Talbot & M. H. Talbot (1965). *Conservation of the Hong Kong Countryside: Summary Report and Recommendation.* Government Printer, Hong Kong.

28. C. Y. Jim & F. Y. Wong (2006). An evaluation of the Country Parks system in Hong Kong since its establishment in 1976. *In:* C. Y. Jim & R. T. Corlett (eds), *Sustainable Management of Protected Areas for Future Generations*, pp. 35–58. Friends of the Country Parks, Hong Kong; World Conservation Union, Gland, Switzerland.

29. H. T. Chang, B. S. Wang, Y. K. Hu, P. X. Bi, Y. H. Chung, Y. Lu & S. X. Yu (1989). The vegetation of Hong Kong. *Acta Scientiarum Naturalium Universitatis Sunyatseni*, supplement 8(2): 1–172.

30. S. L. Thrower (1970). Floristics of the fung shui wood. *In:* L. B. Thrower (ed.), *The Vegetation of Hong Kong: Its Structure and Change*, pp. 57–63. Royal Asiatic Society, Hong Kong Branch, Hong Kong.

31. W. H. Chu & F. W. Xing (1997). A checklist of vascular plants found in *fung shui* woods in Hong Kong. *Memoirs of the Hong Kong Natural History Society* 21: 151–171.

32. P. Kenrick & P. R. Crane (1997). *The Origin and Early Diversification of Land Plants.* Smithsonian Institution Press, Washington.

33. D. E. Soltis, P. S. Soltis, P. K. Endress & M. W. Chase (2005). *Phylogeny and Evolution of Angiosperms.* Sinauer Associates, Sunderland, Massachusetts.

34. N.-H. Xia (2007). Gymnosperms. *In:* Hong Kong Herbarium & South China Botanical Garden (eds), *Flora of Hong Kong.* Vol. 1, pp. 1–15. Agriculture, Fisheries & Conservation Department, Hong Kong.

35. M. Proctor, P. Yeo & A. Lack (1996). *The Natural History of Pollination.* Harper-Collins, London.

36. R. T. Corlett (2004). Flower visitors and pollination in the Oriental (Indomalayan) Region. *Biological Reviews* 79: 497–532.

37. S. Hu, D. L. Dilcher, D. M. Jarzen & D. W. Taylor (2008). Early steps of angiosperm-pollinator coevolution. *Proceedings of the National Academy of Sciences of the United States of America* 105: 240–245.

38. R. W. Spjut (1994). A systematic treatment of fruit types. *Memoirs of the New York Botanical Garden* 70: 1–182.

39. R. T. Corlett (1998). Frugivory and seed dispersal by vertebrates in the Oriental (Indomalayan) Region. *Biological Reviews* 73: 413–448.

40. R. T. Corlett (1998). Frugivory and seed dispersal by birds in Hong Kong shrubland. *Forktail* 13: 23–27.

41. J. McNeill, F. R. Barrie, W. R. Buck, V. Demoulin, W. Greuter, D. L. Hawksworth, P. S. Herendeen, S. Knapp, K. Marhold, J. Prado, W. F. Prud'homme van Reine, G. F. Smith, J. H. Wiersema & N. J. Turland, eds (2012). International code of nomenclature for algae, fungi, and plants. *Regnum Vegetabile* 154: 1–208.

42. S. Sherwood (2005). *A New Flowering: 1000 Years of Botanical Art.* Ashmolean Museum, University of Oxford, Oxford.

43. A. R. Arber (1912). *Herbals: Their Origin and Evolution.* Cambridge University Press, Cambridge.

44. R. Desmond (1987). *A Celebration of Flowers: Two Hundred Years of Curtis's Botanical Magazine.* Royal Botanic Gardens, Kew, London.

45. D.-L. Wu (2008). Mimosaceae. *In:* Hong Kong Herbarium & South China Botanical Garden (eds), *Flora of Hong Kong.* Vol. 2, pp. 36–46. Agriculture, Fisheries & Conservation Department, Hong Kong.

46. N. H. Boke (1940). Histogenesis and morphology of the phyllode in certain species of *Acacia. American Journal of Botany* 27: 73–90.

47. V. H. Boughton (1986). Phyllode structure, taxonomy and distribution in some Australian acacias. *Australian Journal of Botany* 34: 663–674.

48. T. Brodribb & R. H. Hill (1993). A physiological comparison of leaves and phyllodes in *Acacia melanoxylon. Australian Journal of Botany* 41: 293–305.

49. M. L. So (2004). The occurrence of extrafloral nectaries in Hong Kong plants. *Botanical Bulletin of Academia Sinica* 45: 237–245.

50. D. J. O'Dowd & A. M. Gill (1986). Seed dispersal syndromes in Australian *Acacia. In:* D. R. Murray (ed.), *Seed Dispersal,* pp. 87–121. Academic Press, Sydney.

51. R. T. Corlett (2005). Interactions between birds, fruit bats and exotic plants in urban Hong Kong, South China. *Urban Ecosystems* 8: 275–283.

52. S.-H. Lin, S.-Y. Chang & S.-H. Chen (1993). The study of bee-collected pollen loads in Nantou, Taiwan. *Taiwania* 38: 117–133.

53. E. W. S. Lee, B. C. H. Hau & R. T. Corlett (2005). Natural regeneration in exotic tree plantations in Hong Kong, China. *Forest Ecology and Management* 212: 358–366.

54. C.-H. Chou, C.-Y. Fu, S.-Y. Li & Y.-F. Wang (1998). Allelopathic potential of *Acacia confusa* and related species in Taiwan. *Journal of Chemical Ecology* 24: 2131–2150.

55. N.-H. Xia (2008). Aceraceae. *In:* Hong Kong Herbarium & South China Botanical Garden (eds), *Flora of Hong Kong.* Vol. 2, pp. 261–263. Agriculture, Fisheries & Conservation Department, Hong Kong.

56. K. Matsui (1991). Pollination ecology of four *Acer* species in Japan with special reference to bee pollinators. *Plant Species Biology* 6: 117–120.

57. M. G. Harrington, K. J. Edwards, S. A. Johnson, M. W. Chase & P. A. Gadek (2005). Phylogenetic inference in Sapindaceae *sensu lato* using plastid *matK* and *rbcL* DNA sequences. *Systematic Botany* 30: 366–382.

58. A. N. Muellner-Riehl, A. Weeks, J. W. Clayton, S. Buerki, L. Nauheimer, Y.-C. Chiang, S. Cody & S. K. Pell (2016). Molecular phylogenetics and molecular clock dating of Sapindales based on plastid *rbcL*, *atpB* and *trnL-trnF* DNA sequences. *Taxon* 65: 1019–1036.

59. S. Buerki, P. P. Lowry II, N. Alvarez, S. G. Razafimandimbison, P. Küpfer & M. W. Callmander (2010). Phylogeny and circumscription of Sapindaceae revisited: molecular sequence data, morphology, and biogeography support recognition of a new family, Xanthoceraceae. *Plant Ecology and Evolution* 143: 148–159.

60. N.-H. Xia (2008). Rutaceae. *In:* Hong Kong Herbarium & South China Botanical Garden (eds), *Flora of Hong Kong.* Vol. 2, pp. 273–286. Agriculture, Fisheries & Conservation Department, Hong Kong.

61. R. T. Corlett (2001). Pollination in a degraded tropical landscape: a Hong Kong case study. *Journal of Tropical Ecology* 17: 155–161.

62. R. T. Corlett (1996). Characteristics of vertebrate-dispersed fruits in Hong Kong. *Journal of Tropical Ecology* 12: 819–833.

63. I. W. P. Ko, R. T. Corlett & R.-J. Xu (1998). Sugar composition of wild fruits in Hong Kong, China. *Journal of Tropical Ecology* 14: 381–387.

64. M. Luckow & J. Grimes (1997). A survey of the anther glands in the mimosoid legume tribes Parkieae and Mimoseae. *American Journal of Botany* 84: 285–297.

65. T. C. de Barros & S. P. Teixeira (2016). Revisited anatomy of anther glands in mimosoids (Leguminosae). *International Journal of Plant Sciences* 177: 18–33.

66. L. van der Pijl (1982). *Principles of Dispersal in Higher Plants.* 3rd ed. Springer-Verlag, Berlin.

67. Q.-M. Hu (2009). Rubiaceae. *In:* Hong Kong Herbarium & South China Botanical Garden (eds), *Flora of Hong Kong.* Vol. 3, pp. 203–240. Agriculture, Fisheries & Conservation Department, Hong Kong.

68. H. N. Ridley (1930). *The Dispersal of Plants Throughout the World.* L. Reeve, Ashford, Kent.

69. Y.-F. Deng & N.-H. Xia (2007). Theaceae. *In:* Hong Kong Herbarium & South China Botanical Garden (eds), *Flora of Hong Kong.* Vol. 1, pp. 178–193. Agriculture, Fisheries & Conservation Department, Hong Kong.

70. T. Min (T.-L. Ming) & B. Bartholomew (2007). Theaceae. *In:* Z. Y. Wu, P. H. Raven & D. Y. Hong (eds), *Flora of China.* Vol. 12, pp. 366–478. Science Press, Beijing; Missouri Botanical Garden, St Louis.

71. J. Schönenberger, A. A. Anderberg & K. J. Sytsma (2005). Molecular phylogenetics and patterns of floral evolution in the Ericales. *International Journal of Plant Sciences* 166: 265–288.

72. X.-S. Yu (2012). Textual Research on Yang Tong (杨桐, *Adinandra millettii*), Hai Tong (海桐, *Pittosporum tobira*) and Chai Tong (拆桐). *Journal of Beijing Forestry University (Social Sciences)* 11: 24–27.

73. N.-H. Xia (2008). Alangiaceae. *In:* Hong Kong Herbarium & South China Botanical Garden (eds), *Flora of Hong Kong.* Vol. 2, p. 164. Agriculture, Fisheries & Conservation Department, Hong Kong.

74. J. V. LaFrankie (2010). *Trees of Tropical Asia: An Illustrated Guide to Diversity.* Black Tree Publications, San Fernando, La Union, Philippines.

75. J. A. Duke & E. S. Ayensu (1985). *Medicinal Plants of China.* Reference Publications, Algonac, Michigan, USA.

76. W. Tang & G. Eisenbrand (1992). *Chinese Drugs of Plant Origin: Chemistry, Pharmacology, and Use in Traditional and Modern Medicine.* Springer-Verlag, Berlin & Heidelberg, Germany.

77. P.-T. Li (2008). Euphorbiaceae. *In:* Hong Kong Herbarium & South China Botanical Garden (eds), *Flora of Hong Kong.* Vol. 2, pp. 194–237. Agriculture, Fisheries & Conservation Department, Hong Kong.

78. H. Krisnawati, M. Kallio & M. Kanninen (2011). *Aleurites moluccana (L.) Willd.: Ecology, Silviculture and Productivity.* Center for International Forestry Research, Bogor, Indonesia.

79. M.-H. Ho (1981). *Hong Kong Poisonous Plants.* The Urban Council, Hong Kong.

80. C. Y. Jim (1990). *Trees in Hong Kong: Species for Landscape Planting.* Hong Kong University Press, Hong Kong.

81. V. Grant (1950). The protection of ovules in flowering plants. *Evolution* 4: 179–201.

82. R.-J. Zhang & J. Schönenberger (2014). Early floral development of Pentaphylacaceae (Ericales) and its systematic implications. *Plant Systematics and Evolution* 300: 1547–1560.

83. R.-J. Wang & R. M. K. Saunders (2007). Annonaceae. *In:* Hong Kong Herbarium & South China Botanical Garden (eds), *Flora of Hong Kong.* Vol. 1, pp. 30–36. Agriculture, Fisheries & Conservation Department, Hong Kong.

84. I. H. Burkill (1935). *A Dictionary of the Economic Products of the Malay Peninsula.* Vol. 1. Crown Agents for the Colonies, London.

85. R. M. K. Saunders (2010). Floral evolution in the Annonaceae: hypotheses of homeotic mutations and functional convergence. *Biological Reviews* 85: 571–591.

86. C.-C. Pang & R. M. K. Saunders (2014). The evolution of alternative mechanisms that promote outcrossing in Annonaceae, a self-compatible family of early-divergent angiosperms. *Botanical Journal of the Linnean Society* 174: 93–109

87. R. M. K. Saunders (2012). The diversity and evolution of pollination systems in Annonaceae. *Botanical Journal of the Linnean Society* 169: 222–244.

88. A. K. van Setten & J. Koek-Noorman (1992). Fruits and seeds of Annonaceae: morphology and its significance for classification and identification. *Bibliotheca Botanica* 142: 1–101 + pl. 1–50.

89. W. Roxburgh (1832). *Flora Indica.* Vol. 3. W. Thacker, Calcutta.

90. A. M. Schot (1995). A synopsis of taxonomic changes in *Aporosa* Blume (Euphorbiaceae). *Blumea* 40: 449–460.

91. A. A. Akers, M. A. Islam & V. Nijman (2013). Habitat characterization of western hoolock gibbons *Hoolock hoolock* by examining home range microhabitat use. *Primates* 54: 341–348.

92. D. S. Hill, P. Hore & I. W. B. Thornton (1982). *Insects of Hong Kong.* Hong Kong University Press, Hong Kong.

93. N.-H. Xia (2008). Thymelaeaceae. *In:* Hong Kong Herbarium & South China Botanical Garden (eds), *Flora of Hong Kong.* Vol. 2, pp. 132–135. Agriculture, Fisheries & Conservation Department, Hong Kong.

94. J. K. L. Yip & P. C. C. Lai (2004). The nationally rare and endangered plant, *Aquilaria sinensis*: its status in Hong Kong. *Hong Kong Biodiversity* 7: 14–16.

95. Y. Liu, H. Chen, Y. Yang, Z. Zhang, J. Wei, H. Meng, W. Chen, J. Feng, B. Gan, X. Chen, Z. Gao, J. Huang, B. Chen & H. Chen (2013). Whole-tree agarwood-inducing technique: an efficient novel technique for producing high-quality agarwood in cultivated *Aquilaria sinensis* trees. *Molecules* 18: 3086–3106.

96. C. Y. Jim (2015). Cross-border itinerant poaching of agarwood in Hong Kong's peri-urban forests. *Urban Forestry and Urban Greening* 14: 420–431.

97. International Union for Conservation of Nature (2015). *The IUCN Red List of Threatened Species.* http://www.iucnredlist.org.

98. South China Institute of Botany & Hong Kong Herbarium (2003). *Rare and Precious Plants of Hong Kong*. Agriculture, Fisheries & Conservation Department, Hong Kong.

99. K.-C. Iu (1983). The cultivation of the "incense tree" (*Aquilaria sinensis*). *Journal of the Royal Asiatic Society, Hong Kong Branch* 23: 247–249.

100. T. Soehartono & A. C. Newton (2001). Reproductive ecology of *Aquilaria* spp. in Indonesia. *Forest Ecology and Management* 152: 59–71.

101. G. Chen, C. Liu & W. Sun (2016). Pollination and seed dispersal of *Aquilaria sinensis* (Lour.) Gilg (Thymelaeaceae): an economic plant species with extremely small populations in China. *Plant Diversity* 38: 227–232.

102. A. Y. S. Ng & B. C. H. Hau (2009). Nodulation of native woody legumes in Hong Kong, China. *Plant and Soil* 316: 35–43.

103. B. C. H. Hau & R. T. Corlett (2002). A survey of trees and shrubs on degraded hillsides in Hong Kong. *Memoirs of the Hong Kong Natural History Society* 25: 83–94.

104. Q.-M. Hu (2007). Myrsinaceae. *In:* Hong Kong Herbarium & South China Botanical Garden (eds), *Flora of Hong Kong*. Vol. 1, pp. 295–305. Agriculture, Fisheries & Conservation Department, Hong Kong.

105. H. Miehe (1911). Die Bakterienknoten an den Blatträndern der *Ardisia crispa* A. DC. *Abhandlungen der Mathematisch-Physischen Classe der Königlich Sächsischen Gesellschaft der Wissenschaften, Leipzig* 32: 399–431.

106. I. M. Miller, I. C. Gardner & A. Scott (1984). Structure and function of trichomes in the shoot tip of *Ardisia crispa* (Thunb.) A. DC. (Myrsinaceae). *Botanical Journal of the Linnean Society* 88: 223–236.

107. I. M. Miller, I. C. Gardner & A. Scott (1983). The development of marginal leaf nodules in *Ardisia crispa* (Thunb.) A. DC. (Myrsinaceae). *Botanical Journal of the Linnean Society* 86: 237–252.

108. I. M. Miller (1990). Bacterial leaf nodule symbiosis. *Advances in Botanical Research* 17: 163–234.

109. D.-L. Wu (2007). Moraceae. *In:* Hong Kong Herbarium & South China Botanical Garden (eds), *Flora of Hong Kong*. Vol. 1, pp. 101–115. Agriculture, Fisheries & Conservation Department, Hong Kong.

110. R. T. Corlett (2011). Seed dispersal in Hong Kong, China: past, present and possible futures. *Integrative Zoology* 6: 97–109.

111. J. W. Purseglove (1968). *Tropical Crops: Dicotyledons*. Vol. 1. Longmans, London.

112. J. Barrow (1831). *The Eventful History of the Mutiny and Piratical Seizure of H. M. S. Bounty: Its Causes and Consequences*. J. Murray, London.

113. C. Alexander (2003). *The Bounty: The True Story of the Mutiny on the Bounty*. Harper Collins, London.

114. N.-H. Xia (2008). Cornaceae. *In:* Hong Kong Herbarium & South China Botanical Garden (eds), *Flora of Hong Kong*. Vol. 2, pp. 165–166. Agriculture, Fisheries & Conservation Department, Hong Kong.

115. R. K. Brummitt (2007). Aucubaceae. *In:* V. H. Heywood, R. K. Brummitt, A. Culham & O. Seberg (eds), *Flowering Plant Families of the World*, pp. 52–53. Firefly Books, Ontario.

116. Q.-Y. Xiang (2016). Aucubaceae. *In:* J. W. Kadereit & V. Bittrich (eds), *The Families and Genera of Vascular Plants*. Vol. 14, pp. 37–40. Springer, Cham, Switzerland.

117. D. E. Soltis, P. S. Soltis, M. W. Chase, M. E. Mort, D. C. Albach, M. Zanis, V. Savolainen, W. H. Hahn, S. B. Hoot, M. F. Fay, M. Axtell, S. M. Swensen, L. M. Price, W. H. Kress, K. C. Nixon & J. S. Farris (2000). Angiosperm phylogeny inferred from 18S rDNA, *rbc*L, and *atp*B sequences. *Botanical Journal of the Linnean Society* 133: 381–461.

118. Q. Zhao, Z. Deng & J. Xu (1991). Natural foods and their ecological implications for *Macaca thibetana* at Mount Emei, China. *Folia Primatologica* 57: 1–15.

119. N.-H. Xia & Y.-F. Deng (2008). Caesalpiniaceae. *In:* Hong Kong Herbarium & South China Botanical Garden (eds), *Flora of Hong Kong*. Vol. 2, pp. 46–59. Agriculture, Fisheries & Conservation Department, Hong Kong.

120. S. T. Dunn (1906). Report on the Botanical and Forestry Department, for the year 1905. *Administration Report [Hong Kong Government]* 1906: 439–452.

121. W. J. Tutcher (1915). Report on the Botanical and Forestry Department for the year 1914. *Administration Report [Hong Kong Government]* 1915: M1–M36.

122. S. T. Dunn (1908). New Chinese plants. *Journal of Botany* 46: 324–326.

123. C. P. Y. Lau, L. Ramsden & R. M. K. Saunders (2005). Hybrid origin of "*Bauhinia blakeana*" (Leguminosae: Caesalpinioideae), inferred using morphological, reproductive, and molecular data. *American Journal of Botany* 92: 525–533.

124. C. Y. Mak, K. S. Cheung, P. Y. Yip & H. S. Kwan (2008). Molecular evidence for the hybrid origin of *Bauhinia blakeana* (Caesalpinioideae). *Journal of Integrative Plant Biology* 50: 111–118.

125. N.-H. Xia (2007). Bombacaceae. *In:* Hong Kong Herbarium & South China Botanical Garden (eds), *Flora of Hong Kong*. Vol. 1, pp. 214–215. Agriculture, Fisheries & Conservation Department, Hong Kong.

126. A. Bhattacharya & S. Mandal (2000). Pollination biology in *Bombax ceiba* Linn. *Current Science* 79: 1706–1712.

127. A. J. S. Raju, S. P. Rao & K. Rangaiah (2005). Pollination by bats and birds in the obligate outcrosser *Bombax ceiba* L. (Bombacaceae), a tropical dry season flowering tree species in the Eastern Ghats forests of India. *Ornithological Science* 4: 81–87.

128. T. A. Davis & K. O. Mariamma (1965). The three kinds of stamens in *Bombax ceiba* L. (Bombacaceae). *Bulletin du Jardin Botanique de l'État, Bruxelles* 35: 185–211.

129. N.-H. Xia (2008). Rhizophoraceae. *In:* Hong Kong Herbarium & South China Botanical Garden (eds), *Flora of Hong Kong*. Vol. 2, pp. 162–164. Agriculture, Fisheries & Conservation Department, Hong Kong.

130. T. Elmqvist & P. A. Cox (1996). The evolution of vivipary in flowering plants. *Oikos* 77: 3–9.

131. P. B. Tomlinson & P. A. Cox (2000). Systematic and functional anatomy of seedlings in mangrove Rhizophoraceae: vivipary explained? *Botanical Journal of the Linnean Society* 134: 215–231.

132. N.-H. Xia (2009). Verbenaceae. *In:* Hong Kong Herbarium & South China Botanical Garden (eds), *Flora of Hong Kong*. Vol. 3, pp. 80–97. Agriculture, Fisheries & Conservation Department, Hong Kong.

133. M. Kato, Y. Kosaka, A Kawakita, Y. Okuyama, C. Kobayashi, T. Phimminith & D. Thongphan (2008). Plant-pollinator interactions in tropical monsoon forests in Southeast Asia. *American Journal of Botany* 95: 1375–1394.

134. Y. Tu, L. Sun, M. Guo & W. Chen (2013). The medicinal uses of *Callicarpa* L. in traditional Chinese medicine: an ethnopharmacological, phytochemical and pharmacological review. *Journal of Ethnopharmacology* 146: 465–481.

135. W. J. Tutcher (1905). Descriptions of some new species, and notes on other Chinese plants. *Journal of the Linnean Society, Botany* 37: 58–70.

136. F. Xing, S.-C. Ng & L. K. C. Chau (2000). Gymnosperms and angiosperms of Hong Kong. *Memoirs of the Hong Kong Natural History Society* 23: 21–135.

137. P. C.-C. Lai & K.-L. Yip (2009). Flora conservation in Hong Kong: efforts to preserve plant diversity. *In:* Hong Kong Herbarium & South China Botanical Garden (eds), *Flora of Hong Kong.* Vol. 3, pp. xvi–xxiv. Agriculture, Fisheries & Conservation Department, Hong Kong.

138. H. Abe, R. Matsuki, S. Ueno, M. Nashimoto & M. Hasegawa (2006). Dispersal of *Camellia japonica* seeds by *Apodemus speciosus* revealed by maternity analysis of plants and behavioral observation of animal vectors. *Ecological Research* 21: 732–740.

139. K. P. S. Chung & R. T. Corlett (2006). Rodent diversity in a highly degraded tropical landscape: Hong Kong, South China. *Biodiversity and Conservation* 15: 4521–4532.

140. B. Seemann (1859). Synopsis of the genera *Camellia* and *Thea. Transactions of the Linnean Society of London* 22: 337–352 + pls 60–61.

141. J. G. Champion (1853). The Ternstrœmiaceous plants of Hong Kong. *Transactions of the Linnean Society of London* 21: 111–116 + pls 12–13.

142. Q.-M. Hu (2007). Fagaceae. *In:* Hong Kong Herbarium & South China Botanical Garden (eds), *Flora of Hong Kong.* Vol. 1, pp. 126–139. Agriculture, Fisheries & Conservation Department, Hong Kong.

143. B. S. Fey & P. K. Endress (1983). Development and morphological interpretation of the cupule in Fagaceae. *Flora* 173: 451–468.

144. L. L. Forman (1966). On the evolution of cupules in the Fagaceae. *Kew Bulletin* 18: 385–419.

145. Z. B. Zhang, Z. S. Xiao & H. J. Li (2005). Impact of small rodents on tree seeds in temperate and subtropical forests, China. *In:* P. M. Forget, J. E. Lambert, P. E. Hulme & S. B. Vander Wall (eds), *Seed Fate: Predation, Dispersal and Seeding Establishment,* pp. 269–282. CABI, Wallingford, UK.

146. N.-H. Xia (2007). Casuarinaceae. *In:* Hong Kong Herbarium & South China Botanical Garden (eds), *Flora of Hong Kong.* Vol. 1, p. 140. Agriculture, Fisheries & Conservation Department, Hong Kong.

147. L. B. Zhang & N. J. Turland (2013). Equisetaceae. *In:* Z. Y. Wu, P. H. Raven & D. Y. Hong (eds), *Flora of China.* Vol. 2–3, pp. 67–72. Science Press, Beijing; Missouri Botanical Garden, St Louis.

148. J. G. Torrey (1976). Initiation and development of root nodules of *Casuarina* (Casuarinaceae). *American Journal of Botany* 63: 335–344.

149. B. Nagarajan, A. Nicodemus, V. Sivakumar, A. K. Mandal, G. Kumaravelu, R. S. C. Jayaraj, V. N. Bai & R. Kamalakannan (2006). Phenology and control pollination studies in *Casuarina equisetifolia* Forst. *Silvae Genetica* 55: 149–155.

150. D.-L. Wu (2007). Ulmaceae. *In:* Hong Kong Herbarium & South China Botanical Garden (eds), *Flora of Hong Kong.* Vol. 1, pp. 96–100. Agriculture, Fisheries & Conservation Department, Hong Kong.

151. P.-T. Li (2009). Apocynaceae. *In:* Hong Kong Herbarium & South China Botanical Garden (eds), *Flora of Hong Kong.* Vol. 3, pp. 16–30. Agriculture, Fisheries & Conservation Department, Hong Kong.

152. D. J. Radford, A. D. Gillies, J. A. Hinds & P. Duffy (1986). Naturally occurring cardiac glycosides. *The Medical Journal of Australia* 144: 540–544.

153. Y. Gaillard, A. Krishnamoorthy & F. Bevalot (2004). *Cerbera adollam*: a 'suicide tree' and cause of death in the state of Kerala, India. *Journal of Ethnopharmacology* 95: 123–126.

154. R. J. Whittaker & S. H. Jones (1994). The role of frugivorous bats and birds in the rebuilding of a tropical forest ecosystem, Krakatau, Indonesia. *Journal of Biogeography* 21: 245–258.

155. H. Nakanishi (1988). Dispersal ecology of maritime plants in the Ryukyu Islands, Japan. *Ecological Research* 3: 163–173.

156. J. M. B. Smith, H. Heatwole, M. Jones & B. M. Waterhouse (1990). Drift disseminules on cays of the Swain Reefs, Great Barrier Reef, Australia. *Journal of Biogeography* 17: 5–17.

157. S. L. Thrower (1988). *Hong Kong Trees.* The Urban Council, Hong Kong.

158. Y. Yang & D.-Z. Fu (2007). Lauraceae. *In:* Hong Kong Herbarium & South China Botanical Garden (eds), *Flora of Hong Kong.* Vol. 1, pp. 37–54. Agriculture, Fisheries & Conservation Department, Hong Kong.

159. A. Moqrich, S. W. Hwang, T. J. Earley, M. J. Petrus, A. N. Murray, K. S. R. Spencer, M. Andahazy, G. M. Story & A. Patapoutian (2005). Impaired thermosensation in mice lacking TRPV3, a heat and camphor sensor in the skin. *Science* 307: 1468–1472.

160. D.-L. Wu (2007). Clusiaceae (Guttiferae). *In:* Hong Kong Herbarium & South China Botanical Garden (eds), *Flora of Hong Kong.* Vol. 1, pp. 196–200. Agriculture, Fisheries & Conservation Department, Hong Kong.

161. N. K. B. Robson (2007). Clusiaceae. *In:* V. H. Heywood, R. K. Brummitt, A. Culham & O. Seberg (eds), *Flowering Plant Families of the World*, pp. 103–104. Firefly Books, Ontario.

162. X. Li, J. Li, N. K. B. Robson & P. F. Stevens (2007). Clusiaceae (Guttiferae). *In:* Z. Y. Wu, P. H. Raven & D. Y. Hong (eds), *Flora of China.* Vol. 13, pp. 1–47. Science Press, Beijing; Missouri Botanical Garden, St Louis.

163. A. Y. Y. Au, R. T. Corlett & B. C. H. Hau (2006). Seed rain into upland plant communities in Hong Kong, China. *Plant Ecology* 186: 13–22.

164. K. C. Nixon (1993). Infrageneric classification of *Quercus* (Fagaceae) and typification of sectional names. *Annales des Sciences Forestières* 50 (supplement 1): 25–34.

165. D.-L. Wu (2008). Fabaceae (Papilionaceae). *In:* Hong Kong Herbarium & South China Botanical Garden (eds), *Flora of Hong Kong.* Vol. 2, pp. 59–119. Agriculture, Fisheries & Conservation Department, Hong Kong.

166. X. Zhang (2011). *Chinese Furniture.* Cambridge University Press, Cambridge.

167. World Conservation Monitoring Centre (1998). *Dalbergia odorifera. The IUCN Red List of Threatened Species* 1998: e.T32398A9698077.

168. D.-L. Wu (2007). Daphniphyllaceae. *In:* Hong Kong Herbarium & South China Botanical Garden (eds), *Flora of Hong Kong.* Vol. 1, pp. 95–96. Agriculture, Fisheries & Conservation Department, Hong Kong.

169. A. M. C. Tang, R. T. Corlett & K. D. Hyde (2005). The persistence of ripe fleshy fruits in the presence and absence of frugivores. *Oecologia* 142: 232–237.

170. J. F. Ma, P. R. Ryan & E. Delhaize (2001). Aluminium tolerance in plants and the complexing role of organic acids. *Trends in Plant Science* 6: 273–278.

171. P. K. Endress (1994). *Diversity and Evolutionary Biology of Tropical Flowers.* Cambridge University Press, Cambridge.

172. J. Léandri (1933). Sur la station d'origine de *Poinciana regia* Boj. *Bulletin du Muséum d'Histoire Naturelle, Paris*, série 2, 5: 413–414.

173. D. J. Du Puy, P. B. Phillipson & R. Rabevohitra (1995). The genus *Delonix* (Leguminosae: Caesalpinioideae: Caesalpinieae) in Madagascar. *Kew Bulletin* 50: 445–475.

174. M. T. K. Arroyo (1981). Breeding systems and pollination biology in Leguminosae. *In:* R. M. Polhill & P. H. Raven (eds), *Advances in Legume Systematics.* Part 2, pp. 723–769. Royal Botanic Gardens, Kew.

175. N.-H. Xia (2008). Sapindaceae. *In:* Hong Kong Herbarium & South China Botanical Garden (eds), *Flora of Hong Kong.* Vol. 2, pp. 257–261. Agriculture, Fisheries & Conservation Department, Hong Kong.

176. W. K. Choo & S. Ketsa (1991). *Dimocarpus longan* Lour. *In:* E. W. M. Verheij & R. E. Coronel (eds), *Plant Resources of South-East Asia. No. 2, Edible Fruits and Nuts,* pp. 146–151. Pudoc, Wageningen.

177. C. A. McConchie, V. Vithanage & D. J. Batten (1994). Intergeneric hybridisation between Litchi (*Litchi chinensis* Sonn.) and Longan (*Dimocarpus longan* Lour.). *Annals of Botany* 74: 111–118.

178. B. Yang, Y. Jiang, J. Shi, F. Chen & M. Ashraf (2011). Extraction and pharmacological properties of bioactive compounds from longan (*Dimocarpus longan* Lour.) fruit—a review. *Food Research International* 44: 1837–1842.

179. N.-H. Xia (2007). Ebenaceae. *In:* Hong Kong Herbarium & South China Botanical Garden (eds), *Flora of Hong Kong.* Vol. 1, pp. 284–286. Agriculture, Fisheries & Conservation Department, Hong Kong.

180. L. S. Contreras & N. R. Lersten (1984). Extrafloral nectaries in Ebenaceae: anatomy, morphology, and distribution. *American Journal of Botany* 71: 865–872.

181. P. H. Wan (2009). *The Role of Masked Palm Civet (*Paguma larvata*) and Small Indian Civet (*Viverricula indica*) in Seed Dispersal in Hong Kong, China.* PhD thesis, The University of Hong Kong, Hong Kong.

182. T. Otani & E. Shibata (2000). Seed dispersal and predation by Yakushima macaques, *Macaca fuscata yakui*, in a warm temperate forest of Yakushima Island, southern Japan. *Ecological Research* 15: 133–144.

183. H.-H. Su & L.-L. Lee (2001). Food habits of Formosan rock macaques (*Macaca cyclopis*) in Jentse, Northeastern Taiwan, assessed by fecal analysis and behavioral observation. *International Journal of Primatology* 22: 359–377.

184. A. Nakamoto, K. Kinjo & M. Izawa (2009). The role of Orii's flying-fox (*Pteropus dasymallus inopinatus*) as a pollinator and a seed disperser on Okinawa-jima Island, the Ryukyu Archipelago, Japan. *Ecological Research* 24: 405–414.

185. Z. Luo & R. Wang (2008). Persimmon in China: domestication and traditional utilizations of genetic resources. *Advances in Horticultural Science* 22: 239–243.

186. I. Soerianegara, D. S. Alonzo, S. Sudo & M. S. M. Sosef (1995). *Diospyros* L. *In:* R. H. M. J. Lemmens, I. Soerianegara & W. C. Wong (eds), *Plant Resources of South-East Asia. No. 5(2), Timber Trees: Minor Commercial Timbers,* pp. 185–205. Backhuys, Leiden.

187. N.-H. Xia (2007). Elaeocarpaceae. *In:* Hong Kong Herbarium & South China Botanical Garden (eds), *Flora of Hong Kong.* Vol. 1, pp. 200–203. Agriculture, Fisheries & Conservation Department, Hong Kong.

188. W. J. Baker, M. J. E. Coode, J. Dransfield, S. Dransfield, M. M. Harley, P. Hoffman & R. J. Johns (1998). Patterns of distribution of Malesian vascular plants. *In:* R. Hall & J. D. Holloway (eds), *Biogeography and Geological Evolution of SE Asia,* pp. 243–258. Backhuys, Leiden.

189. S. A. Guerrero & P. C. van Welzen (2011). Revision of Malesian *Endospermum* (Euphorbiaceae) with notes on phylogeny and historical biogeography. *Edinburgh Journal of Botany* 68: 443–482.

190. J. Schaeffer (1971). Revision of the genus *Endospermum* Bth. (Euphorbiaceae). *Blumea* 19: 171–192.

191. N.-H. Xia (2007). Juglandaceae. *In:* Hong Kong Herbarium & South China Botanical Garden (eds), *Flora of Hong Kong.* Vol. 1, pp. 124–125. Agriculture, Fisheries & Conservation Department, Hong Kong.

192. P. S. Manos & D. E. Stone (2001). Evolution, phylogeny, and systematics of the Juglandaceae. *Annals of the Missouri Botanical Garden* 88: 231–269.

193. S.-C. Ng (2007). Ericaceae. *In:* Hong Kong Herbarium & South China Botanical Garden (eds), *Flora of Hong Kong.* Vol. 1, pp. 272–281. Agriculture, Fisheries & Conservation Department, Hong Kong.

194. P. M. Hermann & B. F. Palser (2000). Stamen development in the Ericaceae. I. Anther wall, microsporogenesis, inversion, and appendages. *American Journal of Botany* 87: 934–957.

195. R. T. Corlett (1993). Reproductive phenology of Hong Kong shrubland. *Journal of Tropical Ecology* 9: 501–510.

196. N.-H. Xia & Y.-F. Deng (2008). Rosaceae. *In:* Hong Kong Herbarium & South China Botanical Garden (eds), *Flora of Hong Kong.* Vol. 2, pp. 19–36. Agriculture, Fisheries & Conservation Department, Hong Kong.

197. N. T. Hiep & E. W. M. Verheij (1991). *Eriobotrya japonica* (Thunb.) Lindley. *In:* E. W. M. Verheij & R. E. Coronel (eds), *Plant Resources of South-East Asia. No. 2. Edible Fruits and Nuts,* pp. 161–164. Pudoc, Wageningen.

198. M. L. Badenese, S. Lin, X. Yang, C. Liu & X. Huang (2009). Loquat (*Eriobotrya* Lindl.). *In:* K. M. Folta & S. E. Gardiner (eds), *Plant Genetics and Genomics: Crops and Models.* Vol. 6, *Genetics and Genomics of Rosaceae,* pp. 525–538. Springer, New York.

199. D. P. Abrol (1988). Ecology and behaviour of three bee species pollinating loquat (*Eriobotrya japonica* Lindley). *Proceedings of the Indian National Science Academy B* 54: 161–163.

200. N.-H. Xia (2007). Hamamelidaceae. *In:* Hong Kong Herbarium & South China Botanical Garden (eds), *Flora of Hong Kong.* Vol. 1, pp. 88–95. Agriculture, Fisheries & Conservation Department, Hong Kong.

201. B. S. Carlsward, W. S. Judd, D. E. Soltis, S. Manchester & P. S. Soltis (2011). Putative morphological synapomorphies of Saxifragales and their major subclades. *Journal of the Botanical Research Institute of Texas* 5: 179–196.

202. J. Li, A. L. Bogle & M. J. Donoghue (1999). Phylogenetic relationships in the Hamamelidoideae inferred from sequences of *trn* non-coding regions of chloroplast DNA. *Harvard Papers in Botany* 4: 343–356.

203. P. K. Endress (1993). Hamamelidaceae. *In:* K. Kubitzki, J. G. Rohwer & V. Bittrich (eds), *The Families and Genera of Vascular Plants.* Vol. 2, pp. 322–331. Springer, Cham, Switzerland.

204. D. S. Hill (1967). *Figs* (Ficus *spp.) of Hong Kong.* Hong Kong University Press, Hong Kong.

205. P. W. Lucas & R. T. Corlett (1998). Seed dispersal by long-tailed macaques. *American Journal of Primatology* 45: 29–44.

206. M. Kato, A. Takimura & A. Kawakita (2003). An obligate pollination mutualism and reciprocal diversification in the tree genus *Glochidion* (Euphorbiaceae). *Proceedings of the National Academy of Sciences of the United States of America* 100: 5264–5267.

207. A. Kawakita & M. Kato (2006). Assessment of the diversity and species specificity of the mutualistic association between *Epicephala* moths and *Glochidion* trees. *Molecular Ecology* 15: 3567–3581.

208. T. Okamoto, A. Kawakita & M. Kato (2007). Interspecific variation of floral scent composition in *Glochidion* and its association with host-specific pollinating seed parasite (*Epicephala*). *Journal of Chemical Ecology* 33: 1065–1081.

209. S.-L. Chen & M. G. Gilbert (1994). Verbenaceae. *In:* Z. Y. Wu & P. H. Raven (eds), *Flora of China*. Vol. 17, pp. 1–49. Science Press, Beijing; Missouri Botanical Garden, St Louis.

210. A. J. Paton, D. Springate, S. Suddee, D. Otieno, R. J. Grayer, M. M. Harley, F. Willis, M. S. J. Simmonds, M. P. Powell & V. Savolainen (2004). Phylogeny and evolution of basils and allies (Ocimeae, Labiatae) based on three plastid DNA regions. *Molecular Phylogenetics and Evolution* 31: 277–299.

211. R. de Kok (2012). A revision of the genus *Gmelina* (Lamiaceae). *Kew Bulletin* 67: 293–329.

212. P. V. Bolstad & K. S. Bawa (1982). Self-incompatibility in *Gmelina arborea* L. (Verbenaceae). *Silvae Genetica* 31: 19–21.

213. D.-L. Wu (2007). Sterculiaceae. *In:* Hong Kong Herbarium & South China Botanical Garden (eds), *Flora of Hong Kong*. Vol. 1, pp. 207–213. Agriculture, Fisheries & Conservation Department, Hong Kong.

214. A. J. G. H. Kostermans (1959). A monograph of the genus *Heritiera* Aiton (Sterculiaceae) (including *Argyrodendron* F. v. M. and *Tarrietia* Bl.). *Reinwardtia* 4: 465–583.

215. S.-Y. Hu (2007). Malvaceae. *In:* Hong Kong Herbarium & South China Botanical Garden (eds), *Flora of Hong Kong*. Vol. 1, pp. 215–232. Agriculture, Fisheries & Conservation Department, Hong Kong.

216. S. Liang, R. C. Zhou, S. S. Dong & S. H. Shi (2008). Adaptation to salinity in mangroves: implications on the evolution of salt-tolerance. *Chinese Science Bulletin* 53: 1708–1715.

217. K. Takayama, T. Kajita, J. Murata & Y. Tateishi (2006). Phylogeography and genetic structure of *Hibiscus tiliaceus*—speciation of a pantropical plant sea-drifted seeds. *Molecular Ecology* 15: 2871–2881.

218. H. Kudoh, R. Shimamura, K. Takayama & D. F. Whigham (2006). Consequences of hydrochory in *Hibiscus*. *Plant Species Biology* 21: 127–133.

219. N.-H. Xia & B.-Q. Xu (2007). Flacourtiaceae. *In:* Hong Kong Herbarium & South China Botanical Garden (eds), *Flora of Hong Kong*. Vol. 1, pp. 237–241. Agriculture, Fisheries & Conservation Department, Hong Kong.

220. Q. Yang & S. Zmarzty (2007). Flacourtiaceae. *In:* Z. Y. Wu, P. H. Raven & D. Y. Hong (eds), *Flora of China*. Vol. 13, pp. 112–137. Science Press, Beijing; Missouri Botanical Garden, St Louis.

221. M. W. Chase, S. Zmarzty, M. D. Lledó, K. J. Wurdack, S. M. Swensen & M. F. Fay (2002). When in doubt, put it in Flacourtiaceae: a molecular phylogenetic analysis based on plastid *rbcL* DNA sequences. *Kew Bulletin* 57: 141–181.

222. S.-Y. Hu (2008). Aquifoliaceae. *In:* Hong Kong Herbarium & South China Botanical Garden (eds), *Flora of Hong Kong*. Vol. 2, pp. 182–190. Agriculture, Fisheries & Conservation Department, Hong Kong.

223. A. C. W. Tsang & R. T. Corlett (2005). Reproductive biology of the *Ilex* species (Aquifoliaceae) in Hong Kong, China. *Canadian Journal of Botany* 83: 1645–1654.

224. C.-Y. Hu (1975). In vitro culture of rudimentary embryos of eleven *Ilex* species. *Journal of the American Society for Horticultural Science* 100: 221–225.

225. C.-Y. Hu, F. Rogalski & C. Ward (1979). Factors maintaining *Ilex* rudimentary embryos in the quiescent state and the ultrastructural changes during *in vitro* activation. *Botanical Gazette* 140: 272–279.

226. N.-H. Xia (2007). Illiciaceae. *In:* Hong Kong Herbarium & South China Botanical Garden (eds), *Flora of Hong Kong.* Vol. 1, pp. 65–66. Agriculture, Fisheries & Conservation Department, Hong Kong.

227. N.-H. Xia & R. M. K. Saunders (2008). Illiciaceae. *In:* Z. Y. Wu, P. H. Raven & D. Y. Hong (eds), *Flora of China.* Vol. 7, pp. 32–38. Science Press, Beijing; Missouri Botanical Garden, St Louis.

228. P. K. Endress (2001). The flowers in extant basal angiosperms and inferences on ancestral flowers. *International Journal of Plant Sciences* 162: 1111–1140.

229. G. Sun, D. L. Dilcher, S. Zheng & Z. Zhou (1998). In search of the first flower: a Jurassic angiosperm, *Archaefructus*, from Northeast China. *Science* 282: 1692–1695.

230. G. Sun, Q. Ji, D. L. Dilcher, S. Zheng, K. C. Nixon & X. Wang (2002). Archaefructaceae, a new basal angiosperm family. *Science* 296: 899–904.

231. J. A. Doyle (2008). Integrating molecular phylogenetic and paleobotanical evidence on origin of the flower. *International Journal of Plant Sciences* 169: 816–843.

232. P. K. Endress & J. A. Doyle (2009). Reconstructing the ancestral angiosperm flower and its initial specializations. *American Journal of Botany* 96: 22–66.

233. N.-H. Xia (2008). Grossulariaceae. *In:* Hong Kong Herbarium & South China Botanical Garden (eds), *Flora of Hong Kong.* Vol. 2, p. 15. Agriculture, Fisheries & Conservation Department, Hong Kong.

234. N.-H. Xia (2008). Hydrangeaceae (incl. Philadelphaceae). *In:* Hong Kong Herbarium & South China Botanical Garden (eds), *Flora of Hong Kong.* Vol. 2, pp. 12–14. Agriculture, Fisheries & Conservation Department, Hong Kong.

235. N.-H. Xia & Y.-F. Deng (2007). Droseraceae. *In:* Hong Kong Herbarium & South China Botanical Garden (eds), *Flora of Hong Kong.* Vol. 1, pp. 234–236. Agriculture, Fisheries & Conservation Department, Hong Kong.

236. M. Fishbein, C. Hibsch-Jetter, D. E. Soltis & L. Hufford (2001). Phylogeny of Saxifragales (angiosperms, eudicots): analysis of a rapid, ancient radiation. *Systematic Biology* 50: 817–847.

237. M. Fishbein & D. E. Soltis (2004). Further resolution of the rapid radiation of Saxifragales (Angiosperms, Eudicots) supported by mixed-model Bayesian analysis. *Systematic Botany* 29: 883–891

238. M. W. Parida & B. Jha (2010). Salt tolerance mechanisms in mangroves: a review. *Trees* 24: 199–217.

239. N. Pi, N. F. Y. Tam, Y. Wu & M. H. Wong (2009). Root anatomy and spatial pattern of radial oxygen loss of eight true mangrove species. *Aquatic Botany* 90: 222–230.

240. M. Yamashiro (1961). Ecological study of *Kandelia candel* (L.) Druce, with special reference to the structure and falling of the seedlings. *Hikobia* 2: 209–214.

241. R. T. Corlett (1992). The naturalized flora of Hong Kong: a comparison with Singapore. *Journal of Biogeography* 19: 421–430.

242. S.-C. Ng & R. T. Corlett (2002). The bad biodiversity: alien plant species in Hong Kong. *Biodiversity Science* 10: 109–118.

243. J. Wen, S. M. Ickert-Bond, Z.-L. Nie & R. Li (2010). Timing and modes of evolution of eastern Asian-North American biogeographic disjunctions in seed plants. *In:* M. Long, H. Gu & Z. Zhou (eds), *Darwin's Heritage Today: Proceedings of the Darwin 200 Beijing International Conference*, pp. 252–269. Higher Education Press, Beijing, China.

244. J. Zachos, M. Pagani, L. Sloan, E. Thomas & K. Billups (2001). Trends, rhythms, and aberration in global climate 65 Ma to present. *Science* 292: 686–693.

245. P. S. Herendeen, P. R. Crane & A. Drinnan (1995). Fagaceous flowers, fruits, and cupules from the Campanian (Late Cretaceous) of central Georgia, USA. *International Journal of Plant Sciences* 156: 93–116.

246. H. J. Sims, P. S. Herendeen & P. R. Crane (1998). New genus of fossil Fagaceae from the Santonian (Late Cretaceous) of central Georgia, USA. *International Journal of Plant Sciences* 159: 391–404.

247. Z. Xiao, Z. Zhang & Y. Wang (2005). Effects of seed size on dispersal distance in five rodent-dispersed fagaceous species. *Acta Oecologica* 28: 221–229.

248. Z. Xiao & Z. Zhang (2006). Nut predation and dispersal of Harland Tanoak *Lithocarpus harlandii* by scatter-hoarding rodents. *Acta Oecologica* 29: 205–213.

249. J. de Loureiro (1790). *Flora Cochinchinensis*. Vol. 2. Academy of Sciences, Lisbon.

250. I. A. Fijridiyanto & N. Murakami (2009). Phylogeny of *Litsea* and related genera (Laureae-Lauraceae) based on analysis of *rpb2* gene sequences. *Journal of Plant Research* 122: 283–298.

251. L.-X. Guo (2011). Arecaceae (Palmae). *In:* Hong Kong Herbarium & South China Botanical Garden (eds), *Flora of Hong Kong*. Vol. 4, pp. 17–27. Agriculture, Fisheries & Conservation Department, Hong Kong.

252. K. H. Tan, A. Zubaid & T. H. Kunz (2000). Fruit dispersal by the Lesser dog-faced fruit bat, *Cynopterus brachyotis* (Muller) (Chiroptera: Pteropodidae). *Malayan Nature Journal* 54: 57–62.

253. C. T. Shek (2006). *A Field Guide to the Terrestrial Mammals of Hong Kong*. Friends of the Country Parks/Cosmos Books, Hong Kong.

254. N.-H. Xia & Y.-B. Guo (2008). Myrtaceae. *In:* Hong Kong Herbarium & South China Botanical Garden (eds), *Flora of Hong Kong*. Vol. 2, pp. 135–147. Agriculture, Fisheries & Conservation Department, Hong Kong.

255. H. K. Kwok & R. T. Corlett (2000). The bird communities of a secondary forest and a *Lophostemon confertus* plantation in Hong Kong, South China. *Forest Ecology and Management* 130: 227–234.

256. X. Zhuang & R. T. Corlett (1996). The conservation status of Hong Kong's tree flora. *Chinese Biodiversity* 4 (supplement): 36–43.

257. B. Fiala & U. Maschwitz (1992). Food bodies and their significance for obligate ant-association in the tree genus *Macaranga* (Euphorbiaceae). *Botanical Journal of the Linnean Society* 110: 61–75.

258. M. Heil, T. Koch, A. Hilpert, B. Fiala, W. Boland & K. E. Linsenmair (2001). Extrafloral nectar production of the ant-associated plant, *Macaranga tanarius*, is an induced, indirect, defensive response elicited by jasmonic acid. *Proceedings of the National Academy of Sciences of the United States of America* 98: 1083–1088.

259. U. Moog, B. Fiala, W. Federle & U. Maschwitz (2002). Thrips pollination of the dioecious ant plant *Macaranga hullettii* (Euphorbiaceae) in Southeast Asia. *American Journal of Botany* 89: 50–59.

260. C. Ishida, M. Kono & S. Sakai (2009). A new pollination system: brood-site pollination by flower bugs in *Macaranga* (Euphorbiaceae). *Annals of Botany* 103: 39–44.

261. H. van der Werff (2001). An annotated key to the genera of Lauraceae in the *Flora Malesiana* region. *Blumea* 46: 125–140.

262. J. G. Rohwer, J. Li, B. Rudolph, S. A. Schmidt, H. van der Werff & H.-W. Li (2009). Is *Persea* (Lauraceae) monophyletic? Evidence from nuclear ribosomal ITS sequences. *Taxon* 58: 1153–1167.

263. L. Li, J. Li, J. G. Rohwer, H. van der Werff, Z.-H. Wang & H.-W. Li (2011). Molecular phylogenetic analysis of the *Persea* group (Lauraceae) and its biogeographic implications on the evolution of tropical and subtropical Amphi-Pacific disjunctions. *American Journal of Botany* 98: 1520–1536.

264. M. F. Willson & M. N. Melampy (1983). The effect of bicolored fruit displays on fruit removal by avian frugivores. *Oikos* 41: 27–31.

265. J. G. Rohwer (2009). The timing of nectar secretion in staminal and staminodial glands in Lauraceae. *Plant Biology* 11: 490–494.

266. N.-H. Xia (2007). Magnoliaceae. *In:* Hong Kong Herbarium & South China Botanical Garden (eds), *Flora of Hong Kong.* Vol. 1, pp. 25–30. Agriculture, Fisheries & Conservation Department, Hong Kong.

267. A. Cronquist (1968). *The Evolution and Classification of Flowering Plants.* Nelson, London.

268. A. L. Tahktajan (1969). *Flowering Plants: Origin and Dispersal,* translated by C. Jeffrey. Oliver & Boyd, Edinburgh.

269. T. S. Elias & A.-C. Sun (1985). Morphology and anatomy of foliar nectaries and associated leaves in *Mallotus* (Euphorbiaceae). *Aliso* 11: 17–25.

270. S. Kitamura, T. Yumoto, P. Poonswad, P. Chuailua, K. Plongmai, T. Maruhashi & N. Noma (2002). Interactions between fleshy fruits and frugivores in a tropical seasonal forest in Thailand. *Oecologia* 133: 559–572.

271. C. E. Turner, T. D. Center, D. W. Burrows & G. R. Buckingham (1998). Ecology and management of *Melaleuca quinquenervia*, an invader of wetlands in Florida, U.S.A. *Wetlands Ecology and Management* 5: 165–178.

272. B. A. Barlow (1988). Patterns of differentiation in tropical species of *Melaleuca* L. (Myrtaceae). *Proceedings of the Ecological Society of Australia* 15: 239–247.

273. J. C. Doran (1999). Cajuput oil. *In:* I. Southwell & R. Lowe (eds), *Tea Tree: The Genus Melaleuca,* pp. 221–233. Harwood, Amsterdam.

274. M. J. Lawes, A. Richards, J. Dathe & J. J. Midgley (2011). Bark thickness determines fire resistance of selected tree species from fire-prone tropical savanna in north Australia. *Plant Ecology* 212: 2057–2069.

275. N.-H. Xia (2007). Tiliaceae. *In:* Hong Kong Herbarium & South China Botanical Garden (eds), *Flora of Hong Kong.* Vol. 1, pp. 203–207. Agriculture, Fisheries & Conservation Department, Hong Kong.

276. A. A. Mar & K. Z. Moe (2014). Plant-pollinator interactions of Bago University campus, Bago Region. *Universities Research Journal [Government of the Republic of the Union of Myanmar Ministry of Education]* 6: 259–274.

277. C.-Y. Zeng, Q.-X. Mei, Y.-Q. Gao, H. Lin, J.-W. Feng & Z.-Y. Ou (2009). Experimental study on the pharmacodynamics in analgesic of water-extract of *Microcos paniculata. Chinese Archives of Traditional Chinese Medicine* 27: 1757–1758.

278. L.-P. Zhang & J.-P. Luo (2008). Overview on pharmaceutical research and clinical application of *Microcos paniculata. Journal of Chinese Medicinal Materials* 31: 935–938.

279. C. Bayer, M. F. Fay, A. Y. de Bruijn, V. Savolainen, C. M. Morton, K. Kubitzki, W. A. Alverson & M. W. Chase (1999). Support for an expanded family concept of Malvaceae within a recircumscribed order Malvales: a combined analysis of plastid *atpB* and *rbcL* DNA sequences. *Botanical Journal of the Linnean Society* 129: 267–303.

280. M. R. Cheek (2007). Sparrmanniaceae. *In:* V. H. Heywood, R. K. Brummitt, A. Culham & O. Seberg (eds), *Flowering Plant Families of the World,* pp. 307–308. Firefly Books, Ontario.

281. N.-H. Xia (2007). Myricaceae. *In:* Hong Kong Herbarium & South China Botanical Garden (eds), *Flora of Hong Kong.* Vol. 1, pp. 125 126. Agriculture, Fisheries & Conservation Department, Hong Kong.

282. R.-Q. Li, Z.-D. Chen, A.-M. Lu, D. E. Soltis, P. S. Soltis & P. S. Manos (2004). Phylogenetic relationships in Fagales based on DNA sequences from three genomes. *International Journal of Plant Sciences* 165: 311–324.

283. T. Hiyoshi, H. Sasakawa & M. Yatazawa (1988). Isolation of *Frankia* strains from root nodules of *Myrica rubra. Soil Science and Plant Nutrition* 34: 107–116.

284. H.-L. Li, W. Wang, P. E. Mortimer, R.-Q. Li, D.-Z. Li, K. D. Hyde, J.-C. Xu, D. E. Soltis & Z.-D. Chen (2015). Large-scale phylogenetic analyses reveal multiple gains of actinorhizal nitrogen-fixing symbioses in angiosperms associated with climate change. *Scientific Reports* 5: art. 14023.

285. Z.-L. Li, S.-L. Zhang & D.-M. Chen (1992). Red bayberry (*Myrica rubra* Sieb. & Zucc.): a valuable evergreen tree fruit for tropical and subtropical areas. *Acta Horticulturae* 321: 112–121.

286. W. J. Hooker & G. A. W. Arnott (1841). *The Botany of Captain Beechey's Voyage.* Henry G. Bohn, London.

287. B.-Q. Xu & N.-H. Xia (2009). Oleaceae. *In:* Hong Kong Herbarium & South China Botanical Garden (eds), *Flora of Hong Kong.* Vol. 3, pp. 124–133. Agriculture, Fisheries & Conservation Department, Hong Kong.

288. S.-J. Pei & G.-W. Hu (2013). Other important economic plants. *In:* D.-Y. Hong & S. Blackmore (eds), *Plants of China: A Companion to the* Flora of China, pp. 383–396. Cambridge University Press, Cambridge.

289. H. Ômura, K. Honda & N. Hayashi (2000). Floral scent of *Osmanthus fragrans* discourages foraging behavior of cabbage butterfly, *Pieris rapae. Journal of Chemical Ecology* 26: 655–666.

290. Q.-M. Hu (2008). Rhamnaceae. *In:* Hong Kong Herbarium & South China Botanical Garden (eds), *Flora of Hong Kong.* Vol. 2, pp. 238–242. Agriculture, Fisheries & Conservation Department, Hong Kong.

291. D. O. Burge & S. R. Manchester (2008). Fruit morphology, fossil history, and biogeography of *Paliurus* (Rhamnaceae). *International Journal of Plant Sciences* 169: 1066–1085.

292. H. Nakanishi (1985). Geobotanical and ecological studies on three semi-mangrove plants in Japan. *Japanese Journal of Ecology* 35: 85–92.

293. D.-L. Wu (2011). Pandanaceae. *In:* Hong Kong Herbarium & South China Botanical Garden (eds), *Flora of Hong Kong.* Vol. 4, pp. 27–29. Agriculture, Fisheries & Conservation Department, Hong Kong.

294. P. A. Cox (1990). Pollination and the evolution of breeding systems in Pandanaceae. *Annals of the Missouri Botanical Garden* 77: 816–840.

295. M. A. B. Lee (1985). The dispersal of *Pandanus tectorius* by the land crab *Cardisoma carnifex. Oikos* 45: 169–173.

296. L. A. Boodle (1923). The bacterial nodules of the Rubiaceae. *Bulletin of Miscellaneous Information* 1923: 346–348.

297. N. Grobbelaar & E. G. Groenewald (1974). Nitrogen fixation by nodulated species of *Pavetta* and *Psychotria. Zeitschrift für Pflanzenphysiologie* 73: 103–108.

298. N. R. Lersten (1975). Colleter types in Rubiaceae, especially in relation to the bacterial leaf nodule symbiosis. *Botanical Journal of the Linnean Society* 71: 311–319.

299. I. von Teichman, P. J. Robbertse & C. F. van der Merwe (1982). Contributions to the floral morphology and embryology of *Pavetta gardeniifolia* A. Rich. Part 2. The ovule and megasporogenesis. *South African Journal of Botany* 1: 22–27.

300. N.-H. Xia (2007). Pentaphylacaceae. *In:* Hong Kong Herbarium & South China Botanical Garden (eds), *Flora of Hong Kong.* Vol. 1, pp. 195–196. Agriculture, Fisheries & Conservation Department, Hong Kong.

301. R. C. Evans & T. A. Dickinson (2005). Floral ontogeny and morphology in *Gillenia* ("Spiraeoideae") and subfamily Maloideae C. Weber (Rosaceae). *International Journal of Plant Sciences* 166: 427–447.

302. Y. Xiang, C.-H. Huang, Y. Hu, J. Wen, S. Li, T. Yi, H. Chen, J. Xiang & H. Ma (2017). Evolution of Rosaceae fruit types based on nuclear phylogeny in the context of geological times and genome duplication. *Molecular Biology and Evolution* 34: 262–281.

303. S. Prasad, R. Chellam, J. Krishnaswamy & S. P. Goyal (2004). Frugivory of *Phyllanthus emblica* at Rajaji National Park, northwest India. *Current Science* 87: 1188–1190.

304. J. F. Morton (1960). The emblic (*Phyllanthus emblica* L.). *Economic Botany* 14: 119–128.

305. K. Jaijoy, N. Soonthornchareonnon, A. Panthong & S. Sireeratawong (2010). Anti-inflammatory and analgesic activities of the water extract from the fruit of *Phyllanthus emblica* Linn. *International Journal of Applied Research in Natural Products* 3: 28–35.

306. A. Ihantola-Vormisto, J. Summanen, H. Kankaanranta, H. Vuorela, Z. M. Asmawi & E. Moilanen (1997). Anti-inflammatory activity of extracts from leaves of *Phyllanthus emblica*. *Planta Medica* 63: 518–524.

307. N.-H. Xia (2007). Pinaceae. *In:* Hong Kong Herbarium & South China Botanical Garden (eds), *Flora of Hong Kong.* Vol. 1, pp. 3–5. Agriculture, Fisheries & Conservation Department, Hong Kong.

308. S. V. Meyen (1984). Basic features of gymnosperm systematics and phylogeny as evidenced by the fossil record. *The Botanical Review* 50: 1–111.

309. A. B. Leslie (2010). Flotation preferentially selects saccate pollen during conifer pollination. *New Phytologist* 188: 273–279.

310. N.-H. Xia & K.-L. Yip (2007). Podocarpaceae. *In:* Hong Kong Herbarium & South China Botanical Garden (eds), *Flora of Hong Kong.* Vol. 1, pp. 9–11. Agriculture, Fisheries & Conservation Department, Hong Kong.

311. P. B. Tomlinson (1994). Functional morphology of saccate pollen in conifers with special reference to Podocarpaceae. *International Journal of Plant Sciences* 155: 699–715.

312. P. B. Tomlinson, J. E. Braggins & J. A. Rattenbury (1991). Pollination drop in relation to cone morphology in Podocarpaceae: a novel reproductive mechanism. *American Journal of Botany* 78: 1289–1303.

313. G. Gelbart & P. von Aderkas (2002). Ovular secretions as part of pollination mechanisms in conifers. *Annals of Forest Science* 59: 345–357.

314. H. Nakanishi (1996). Fruit color and fruit size of bird-disseminated plants in Japan. *Vegetatio* 123: 207–218.

315. C. P. Y. Lau & L. Ramsden (2011). Floral biology and breeding system of *Polyspora axillaris* (*Gordonia axillaris*). *Memoirs of the Hong Kong Natural History Society* 27: 83–94.

316. X.-Y. Zhuang & R. T. Corlett (2000). Survival and growth of native tree seedlings in secondary forest of Hong Kong. *Journal of Tropical and Subtropical Botany* 8: 291–300.

317. B. C. H. Hau & K. Y. So (2002). Using native tree species to restore degraded hillsides in Hong Kong, China. *In:* H. C. Sim, S. Appanah & P. B. Durst (eds), *Proceedings of an International Conference on Bringing Back the Forests: Policies and Practices for Degraded Lands and Forests, Kuala Lumpur, Malaysia, 7–10 October 2002*, pp. 179–190. Food and Agriculture Organization of the United Nations, Bangkok.

318. M. A. Vincent (2005). On the spread and current distribution of *Pyrus calleryana* in the United States. *Castanea* 70: 20–31.

319. J. Decaisne (1872). *Le Jardin Fruitier du Museum.* Vol. 1, tab. 8. Firmin Didot, Paris.

320. T. M. Culley & N. A. Hardiman (2009). The role of intraspecific hybridization in the evolution of invasiveness: a case study of the commercial pear tree *Pyrus calleryana. Biological Invasions* 11: 1107–1119.

321. N. A. Hardiman & T. M. Culley (2010). Reproductive success of cultivated *Pyrus calleryana* (Rosaceae) and establishment ability of invasive, hybrid progeny. *American Journal of Botany* 97: 1698–1706.

322. K. A. Schierenbeck & N. C. Ellstrand (2009). Hybridization and the evolution of invasiveness in plants and other organisms. *Biological Invasions* 11: 1093–1105.

323. J. Lindley (1827). An account of a new genus of plants called *Reevesia. The Quarterly Journal of Science, Literature, and Art* 2(2): 109–112.

324. F.-T. Fan (2004). *British Naturalists in Qing China: Science, Empire, and Cultural Encounter.* Harvard University Press, Cambridge, Massachusetts.

325. K. C. Chau (1994). *The Ecology of Fire in Hong Kong.* PhD thesis, The University of Hong Kong, Hong Kong.

326. L. M. Marafa & K. C. Chau (1999). Effect of hill fire on upland soil in Hong Kong. *Forest Ecology and Management* 120: 97–104.

327. L. Gu, Z. Luo, D. Zhang & S. S. Renner (2010). Passerine pollination of *Rhodoleia championii* (Hamamelidaceae) in subtropical China. *Biotropica* 42: 336–341.

328. W. J. Hooker (1850). *Rhodoleia championi.* Capt. Champion's Rhodoleia. *Curtis's Botanical Magazine*, series 3, 6: tab. 4509.

329. N.-H. Xia (2008). Anacardiaceae. *In:* Hong Kong Herbarium & South China Botanical Garden (eds), *Flora of Hong Kong.* Vol. 2, pp. 264–267. Agriculture, Fisheries & Conservation Department, Hong Kong.

330. S. X. Shao, Z. X. Yang & X. M. Chen (2013). Gall development and clone dynamics of the galling aphid *Schlechtendalia chinensis* (Hemiptera: Pemphigidae). *Journal of Economic Entomology* 106: 1628–1637.

331. P. Liu, Z. X. Yang, X. M. Chen & R. G. Foottit (2014). The effect of the gall-forming aphid *Schlechtendalia chinensis* (Hemiptera: Aphididae) on leaf wing ontogenesis in *Rhus chinensis* (Sapindales: Anacardiaceae). *Annals of the Entomological Society of America* 107: 242–250.

332. S. Aggarwal (2001). *Rhus* L. *In:* J. L. C. H. van Valkenburg & N. Bunyapraphatsara (eds), *Plant Resources of South-East Asia. No. 12(2), Medicinal and Poisonous Plants*, pp. 469–474. Backhuys, Leiden.

333. O. Djakpo & W. Yao (2010). *Rhus chinensis* and *Galla chinensis*—folklore to modern evidence: review. *Phytotherapy Research* 24: 1739–1747.

334. X. Li, X. Yin & S. He (2001). Seed dispersal by frugivorous birds in Nanjing Botanical Garden Mem. Sun Yat-Sen in autumn and winter. *Chinese Biodiversity* 9: 68–72.

335. R. Govaerts, D. G. Frodin & A. Radcliffe-Smith (2000). *World Checklist and Bibliography of Euphorbiaceae (and Pandaceae)*. The Board and Trustees of the Royal Botanic Gardens, Kew.

336. N.-H. Xia (2007). Actinidiaceae. *In:* Hong Kong Herbarium & South China Botanical Garden (eds), *Flora of Hong Kong*. Vol. 1, pp. 194–195. Agriculture, Fisheries & Conservation Department, Hong Kong.

337. K. Momose, T. Yumoto, T. Nagamitsu, M. Kato, H. Nagamasu, S. Sakai, R. D. Harrison, T. Itioka, A. A. Hamid & T. Inoue (1998). Pollination biology in a lowland dipterocarp forest in Sarawak, Malaysia. I. Characteristics of the plant-pollinator community in a lowland dipterocarp forest. *American Journal of Botany* 85: 1477–1501.

338. W. A. Haber & K. S. Bawa (1984). Evolution of dioecy in *Saurauia* (Dilleniaceae). *Annals of the Missouri Botanical Garden* 71: 289–293.

339. J. H. Cane (1993). Reproductive role of sterile pollen in *Saurauia* (Actinidiaceae), a cryptically dioecious Neotropical tree. *Biotropica* 25: 493–495.

340. Y.-F. Deng (2008). Araliaceae. *In:* Hong Kong Herbarium & South China Botanical Garden (eds), *Flora of Hong Kong*. Vol. 2, pp. 291–296. Agriculture, Fisheries & Conservation Department, Hong Kong.

341. N. Pei, Z. Luo, M. A. Schlessman & D. Zhang (2011). Synchronized protandry and hermaphroditism in a tropical secondary forest tree, *Schefflera heptaphylla* (Araliaceae). *Plant Systematics and Evolution* 296: 29–39.

342. G. Gardner (1849). Descriptions of some new genera and species of plants, collected in the Island of Hong-Kong by Capt. J. G. Champion, 95th Regt. *Hooker's Journal of Botany and Kew Garden Miscellany* 1: 240–246.

343. N.-H. Xia (2007). Sapotaceae. *In:* Hong Kong Herbarium & South China Botanical Garden (eds), *Flora of Hong Kong*. Vol. 1, pp. 281–283. Agriculture, Fisheries & Conservation Department, Hong Kong.

344. T. D. Pennington (2004). Sapotaceae. *In:* K. Kubitzki (ed.), *The Families and Genera of Vascular Plants*. Vol. 6, pp. 390–421. Springer, Berlin.

345. J. A. Tully (2011). *The Devil's Milk: A Social History of Rubber*. New York University Press, New York.

346. J. P. Mathews (2009). *Chicle: The Chewing Gum of the Americas, from the Ancient Maya to William Wrigley*. The University of Arizona Press, Tucson, Arizona.

347. S. A. Temple (1977). Plant-animal mutualism: coevolution with dodo leads to near extinction of plant. *Science* 197: 885–886.

348. J. B. Iverson (1987). Tortoises, not dodos, and the Tambalacoque tree. *Journal of Herpetology* 21: 229–230.

349. M. J. E. Coode (1983). A conspectus of *Sloanea* (Elaeocarpaceae) in the Old World. *Kew Bulletin* 38: 347–427.

350. S. Kitamura, S. Suzuki, T. Yumoto, P. Chuailua, K. Plongmai, P. Poonswad, N. Noma, T. Maruhashi & C. Suckasam (2005). A botanical inventory of a tropical seasonal forest in Khao Yai National Park, Thailand: implications for fruit-frugivore interactions. *Biodiversity and Conservation* 14: 1241–1262.

351. E. J. H. Corner (1949). The durian theory or the origin of the modern tree. *Annals of Botany* 13: 367–414.

352. N.-H. Xia & Y.-F. Deng (2007). Styracaceae. *In:* Hong Kong Herbarium & South China Botanical Garden (eds), *Flora of Hong Kong.* Vol. 1, pp. 286–289. Agriculture, Fisheries & Conservation Department, Hong Kong.

353. F.-W. Xing (2007). Symplocaceae. *In:* Hong Kong Herbarium & South China Botanical Garden (eds), *Flora of Hong Kong.* Vol. 1, pp. 290–294. Agriculture, Fisheries & Conservation Department, Hong Kong.

354. J. L. M. Aranha Filho, P. W. Fritsch, F. Almeda & A. B. Martins (2009). Cryptic dioecy is widespread in South American species of *Symplocos* section *Barberina* (Symplocaceae). *Plant Systematics and Evolution* 277: 99–104.

355. Y.-C. Wang & J.-M. Hu (2011). Cryptic dioecy of *Symplocos wikstroemiifolia* Hayata (Symplocaceae) in Taiwan. *Botanical Studies* 52: 479–491.

356. T. G. van Lingen (1991). *Syzygium jambos* (L.) Alston. *In:* E. W. M. Verheij & R. E. Coronel (eds), *Plant Resources of South-East Asia. No. 2. Edible Fruits and Nuts,* pp. 296–298. Pudoc, Wageningen.

357. G. P. C. Leung, B. C. H. Hau & R. T. Corlett (2009). Exotic plant invasion in the highly degraded upland landscape of Hong Kong, China. *Biodiversity and Conservation* 18: 191–202.

358. M. Flanagan (2008). Notes on the genus *Tetradium. Curtis's Botanical Magazine* 5: 181–191.

359. M. R. Weiss (1995). Floral color change: a widespread functional convergence. *American Journal of Botany* 82: 167–185.

360. Z.-L. Nie, H. Sun, Y. Meng & J. Wen (2009). Phylogenetic analysis of *Toxicodendron* (Anacardiaceae) and its biogeographic implications on the evolution of north temperate and tropical intercontinental disjunctions. *Journal of Systematics and Evolution* 47: 416–430.

361. Y.-M. Lin, F.-C. Chen & K.-H. Lee (1989). Hinokiflavone, a cytotoxic principle from *Rhus succedanea* and the cytotoxicity of the related biflavonoids. *Planta Medica* 55: 166–168.

362. D. G. Barceloux (2008). *Medical Toxicology of Natural Substances: Foods, Fungi, Medicinal Herbs, Plants, and Venomous Animals.* Wiley, Hoboken, New Jersey.

363. M. O. Tucker & C. R. Swan (1998). The mango-poison ivy connection. *The New England Journal of Medicine* 339: 235.

364. O. Vogl (2000). Oriental lacquer, poison ivy and drying oils. *Journal of Polymer Science: Part A: Polymer Chemistry* 38: 4327–4335.

365. E. W. S. Lee, B. C. H. Hau & R. T. Corlett (2008). Seed rain and natural regeneration in *Lophostemon confertus* plantations in Hong Kong, China. *New Forests* 35: 119–130.

366. M.-Q. Yang, R. van Velzen, F. T. Bakker, A. Sattarian, D.-Z. Li & T.-S. Yi (2013). Molecular phylogenetics and character evolution of Cannabaceae. *Taxon* 62: 473–485.

367. N.-H. Xia (2008). Staphyleaceae. *In:* Hong Kong Herbarium & South China Botanical Garden (eds), *Flora of Hong Kong.* Vol. 2, pp. 256–257. Agriculture, Fisheries & Conservation Department, Hong Kong.

368. A. J. Harris, P.-T. Chen, X.-W. Xu, J.-Q. Zhang, X. Yang & J. Wen (2017). A molecular phylogeny of Staphyleaceae: implications for generic delimitation and classical biogeographic disjunctions in the family. *Journal of Systematics and Evolution* 55: 124–141.

369. S. L. Simmons (2007). Staphyleaceae. *In:* K. Kubitzki (ed.), *The Families and Genera of Vascular Plants.* Vol. 9, pp. 440–445. Springer, Heidelberg.

370. D. Fairchild (1913). The Chinese wood-oil tree. *U. S. Department of Agriculture Bureau of Plant Industry Circular* 108: 1–7 + pl. 1–3.

371. Y.-H. Chen, J.-H. Chen, C.-Y. Chang & C.-C. Chang (2010). Biodiesel production from tung (*Vernicia montana*) oil and its blending properties in different fatty acid compositions. *Bioresource Technology* 101: 9521–9526.

372. W. Stuppy, P. C. van Welzen, P. Klinratana & M. C. T. Posa (1999). Revision of the genera *Aleurites*, *Reutealis* and *Vernicia* (Euphorbiaceae). *Blumea* 44: 73–98.

373. D.-L. Wu (2009). Caprifoliaceae. *In:* Hong Kong Herbarium & South China Botanical Garden (eds), *Flora of Hong Kong*. Vol. 3, pp. 240–244. Agriculture, Fisheries & Conservation Department, Hong Kong.

374. B. C. H. Hau (2000). Promoting native tree species in land rehabilitation in Hong Kong, China. *In:* S. Elliott, J. Kerby, D. Blakesley, D. Hardwick, K. Woods & V. Anusarnsunthorn (eds), *Forest Restoration for Wildlife Conservation*, pp. 109–120. International Tropical Timber Organization and The Forest Restoration Research Unit, Chiang Mai University, Chiang Mai, Thailand.

375. C. C. Baskin, C.-T. Chien, S.-Y. Chen & J. M. Baskin (2008). Germination of *Viburnum odoratissimum* seeds: a new level of morphophysiological dormancy. *Seed Science Research* 18: 179–184.

376. O. Singh & V. Rattan (2013). Allelopathic effects of *Viburnum nervosum* on seed germination and seedling growth of *Abies pindrow* Spach. *Allelopathy Journal* 32: 113–122.

377. K. Kawazu (1980). Isolation of Vibsanines A, B, C, D, E and F from *Viburnum odoratissimum*. *Agricultural and Biological Chemistry* 44: 1367–1372.

378. J. Takeda (1994). Plant phenology, animal behaviour and food-gathering by the coastal people of the Ryukyu Archipelago. *Humans and Nature* 3: 117–137.

379. B.-Q. Xu & N.-H. Xia (2009). Oleaceae. *In:* Hong Kong Herbarium & South China Botanical Garden (eds), *Flora of Hong Kong*. Vol. 3, pp. 124–133. Agriculture, Fisheries & Conservation Department, Hong Kong.

380. J.-Y. Cho, T.-L. Hwang, T.-H. Chang, Y.-P. Lim, P.-J. Sung, T.-H. Lee & J.-J. Chen (2012). New coumarins and anti-inflammatory constituents from *Zanthoxylum avicennae*. *Food Chemistry* 135: 17–23.

——— ❦ ———

Autumn leaves of *Rhus chinensis*

Index of Plant Names

——— ❧ ———

Main entry for each species is shown in **boldface**.

Abacus plant (see Hong Kong abacus plant)

Abarema Pittier

    lucida (Benth.) Kosterm.   50

Acacia (see Ear-leaved Acacia, or Taiwan
    Acacia)

Acacia Miller

    auriculiformis Cunn. ex Benth.   **24–25**, 26

    confusa Merr.   9, 19, 24, **26–27**, 182, 288

    richii auct. non A. Gray   26

Acer L.   28

    lanceolatum auct. non Molliard   28

    oblongum auct. non Wall. ex DC.   28

    sino-oblongum F. P. Metcalf   10, **28–29**

Aceraceae   28

Acronychia   30

Acronychia J. R. Forst. & G. Forst.

    laurifolia Blume   30

    pedunculata (L.) Miq.   16, **30–31**

Actinidiaceae   204

Adenanthera L.   32

    microsperma Teijsm. & Binn. **32–33**, 50, 164

    pavonina auct. non L.   32

        var. *microsperma* (Teijsm. & Binn.) I. C.
            Nielsen   32

Adina (see Pilular Adina)

Adina Salisb.

    globiflora Salisb.   34

    pilulifera (Lam.) Franch. ex Drake
                       34–35, 174

Adinandra (see Millett's Adinandra)

Adinandra W. Jack

    millettii (Hook. & Arn.) Benth. & Hook. f.
        ex Hance   **36–37**, 176

Adoxaceae   238

Agarwood   48

Alangiaceae   38

Alangium (see Chinese Alangium)

Alangium Lam.   20

    chinense (Lour.) Harms   9, 19, **38–39**

Alder   46

Aleurites J. R. Forst. & G. Forst.

    cordata auct. non (Thunb.) R. Br. ex Steud.
                       236

    moluccana (L.) Willd.   **40–41**, 98, 202, 236

    montana (Lour.) E. H. Wilson   236

Alnus Mill.   46

    dioica Roxb.   46

Altingiaceae   132

Amborella Baill.   152

Amborellaceae   152

Amentotaxus Pilg.

    argotaenia (Hance) Pilg.   17

Anacardiaceae   200, 230

Anneslea (see Hainan Anneslea)

Anneslea Wall.

    fragrans Wall.   42, 176, 178

        var. *hainanensis* Kobuski   **42–43**

    hainanensis (Kobuski) Hu   42

Annona L.

    squamosa L.   **44–45**

Annonaceae   44

Apea ear-ring (see Chinese Apea ear-ring)

Apocynaceae   78

Aporosa   46

Aporosa Blume

    chinensis (Champ. & Benth.) Merr.   46

    dioica (Roxb.) Müll. Arg.   **46–47**

    frutescens auct. non Blume   46

    leptostachya Benth.   46

Aporusa (see Aporosa)

Aporusa Blume (see *Aporosa*)

Aquifoliaceae   122

Aquilaria Lam.

    grandiflora Benth.   48

    sinensis (Lour.) Spreng.   16, **48–49**

Araliaceae   206

Archaefructus G. Sun, D. L. Dilcher, S. Zheng &
    Z. Zhou   124

Archidendron F. Muell.
  lucidum (Benth.) I. C. Nielsen
                    50–51, 52, 164
Ardisia (see Asiatic Ardisia)
Ardisia Swartz   52
  crenata Sims   52
  pauciflora Heyne ex Roxb.   52
  quinquegona Blume   **52–53**
Arecaceae   17, 140
Artocarpus (see Silver-back Artocarpus, or
  Sweet Artocarpus)
Artocarpus J. R. Forst. & G. Forst.   54
  altilis (Parkinson) Fosberg   54
  heterophyllus Lam.   54
  hypargyreus Hance ex Benth.   **54–55**
Ash   240
Asiatic Ardisia   52
Aucuba (see Chinese Aucuba)
Aucuba Thunb.   56
  chinensis Benth.   **56–57**
  japonica Thunb.   56
Aucubaceae   56
Avocado   148
Bast (see Cuban bast)
Bauhinia L.   20, 58
  'Blakeana' (see Bauhinia purpurea ×
    variegata 'Blakeana')
  blakeana Dunn   58
  purpurea L.   20, 58
  purpurea × variegata 'Blakeana'
                    9, 20, **58–59**
  variegata L.   20, 58
Bentham's Photinia   178
Betulaceae   46
Bombax L.   60
  ceiba L.   8, 9, 19, 20, **60–63**
  malabaricum DC.   20, 60
Bottle-brush   156
Box (see Brisbane or Brush box)
Brazilian rosewood   86
Breadfruit   54
Brisbane box   144
Bruguiera Savigny
  gymnorhiza (L.) Savigny   **64–65**
Brush box   144
Bucklandia R. Br. ex Griff.

tonkinensis (Lec.) Steenis   106
Buddhist pine   186
Buttonbush (see Chinese buttonbush)
Callery pear   192
Callicarpa   66
Callicarpa L.   66
  nudiflora Hook. & Arn.   **66–67**
  reevesii Wall. ex Schauer   66
Callistemon R. Br.   156
Camellia (see Crapnell's Camellia, False
    Camellia, or Hong Kong Camellia)
Camellia L.   68, 70, 190
  axillaris Roxb. ex Ker Gawl.   188
  crapnelliana Tutch.   **68–69**, 70, 190
  hongkongensis Seem.   **70–71**, 190
  japonica L.   70
  reticulata auct. non Lindl.   190
  spectabilis Champ.   190
Camphor tree   80
Candlenut tree   40
Cannabaceae   76, 232
Cannabis   232
Cannabis L.   232
Carpinus L.
  insularis N. H. Xia, K. S. Pang & Y. H. Tong
                                      7
Carya Nutt.   100
Castanopsis (D. Don) Spach   72
  fissa (Champ. ex Benth.) Rehder & E. H.
    Wilson   10, 68, **72–73**, 249
Casuarina L.
  equisetifolia L.   15, 52, **74–75**, 162
Casuarinaceae   74
Ceiba Miller
  pentandra (L.) Gaertn.   60
Celtis L.
  sinensis Pers.   **76–77**
  tetrandra Roxb.
    subsp. sinensis (Pers.) Y. C. Tang   76
Cephalotaxaceae   17
Cerbera   78
Cerbera L.   78
  manghas L.   **78–79**
  odollam auct. non Gaertn.   78
Champion's oak   84
Chekiang Machilus   148

Chestnut oak	72
Chicle	212
Chinese Alangium	38
Chinese Apea ear-ring	50
Chinese Aucuba	56
Chinese buttonbush	34
Chinese date plum	94
Chinese Elaeocarpus	96
Chinese fan palm	140
Chinese guger tree	208
Chinese hackberry	76
Chinese holly	122
Chinese Liquidambar	132
Chinese nettle tree	76
Chinese New Year flower	102
Chinese red pine	14, 184
Chinese Scolopia	210
Chinese Sloanea	214
Cinnamomum Trew	
camphora (L.) J. Presl	16, **80–81**, 136
Clusiaceae	82, 110
Coastal Heritiera	116
Common Pentaphylax	176
Common Tutcheria	190
Congested Symplocos	220
Cork-leaved snow-bell	218
Cornaceae	56
Cotton tree	8, 60
Cow wood (see Yellow cow wood)	
Crapnell's Camellia	68
Crataegus L.	
indica L.	196
Cratoxylum Blume	
cochinchinense (Lour.) Blume	**82–83**, 110
ligustrinum (Spach) Blume	82
polyanthum Korth.	82
Cuban bast	118
Cyclobalanopsis Oerst.	84
championii (Benth.) Oerst.	**84–85**, 134
Cyminosma Gaertn.	
pedunculata DC.	30
Dalbergia L. f.	86
assamica Benth.	**86–87**
balansae Prain	86
nigra (Vell.) Allem. ex Benth.	86
odorifera T. Chen	86
sissoo Roxb. ex DC.	86
Dalrympelea Roxb.	234
Daphniphyllaceae	88
Daphniphyllum	88
Daphniphyllum Blume	88
calycinum Benth.	**88–89**
Date plum (see Chinese or Japanese date plum)	
Delonix Raf.	
regia (Bojer) Raf.	10, 86, **90–91**
Dense-flowered sweet-leaf	220
Dichroa Lour.	126
Dimocarpus Lour.	
longan Lour.	9, 10, 16, **92–93**
Diospyros L.	94
kaki Thunb.	94
morrisiana Hance	**94–95**
Disanthus Maxim.	106
Dragon's eye	92
Drosera L.	126
Droseraceae	126
Duckfoot tree	206
Ear-leaved Acacia	24
Ear-pod wattle	24
Ebenaceae	94
Ebony	94
Echinocarpus Blume	
sinensis Hance	214
Elaeocarpaceae	96, 214
Elaeocarpus (see Chinese Elaeocarpus)	
Elaeocarpus L.	96
chinensis (Gardner & Champ.) Hook. f. & Benth.	**96–97**, 214
Elephant's ear	146
Elm	232
Emarginate-leaved Ormosia	164
Emblic	180
Endospermum	98
Endospermum Benth.	98
chinense Benth.	40, **98–99**, 202
moluccanum (Teijsm. & Binn.) Kurz	98
myrmecophilum L. S. Sm.	98
Engelhardia (see Roxburgh's Engelhardia)	
Engelhardia Leschen. ex Blume	
chrysolepis Hance	100
roxburghiana Lindl. ex Wall.	**100–101**
wallichiana auct. non Lindl.	100

Enkianthus	102
Enkianthus Lour.	
quinqueflorus Lour.	**102–103**, 281
Equisetum L.	74
Ericaceae	102
Eriobotrya Lindl.	
japonica (Thunb.) Lindl.	**104–105**, 196
Eucalyptus L'Hér.	
robusta Sm.	15
Eucommiaceae	56
Eugenia L.	
jambos L.	222
Euodia J. R. Forst. & G. Forst.	224
'meliaefolia' (Hance ex Walp.) Benth.	224
meliifolia (Hance ex Walp.) Benth.	224
Euphorbiaceae 16, 40, 46, 98, 146, 154, 202, 236	
Euphoria Comm. ex Juss.	
longan (Lour.) Steud.	92
Euscaphis Siebold & Zucc.	234
Evodia (see Melia-leaved Evodia)	
Evodia J. R. Forst. & G. Forst. (see *Euodia*)	
Exbucklandia	106
Exbucklandia R. W. Br.	106
tonkinensis (Lec.) H. T. Chang	**106–107**
Fabaceae 20, 24, 26, 32, 50, 58, 86, 90, 130, 164	
Fagaceae	14, 16, 68, 72, 84, 134
False Camellia	188
Fan palm (see Chinese fan palm)	
Ficus L.	108
laceratifolia auct. non Lévl. & Van.	54
variolosa Lindl. ex Benth.	
	10, 54, **108–109**, 112
Fig (see Mountain fig)	
Flacourtiaceae	120, 210
Flame-of-the-forest	90
Formosan gum	132
Fountain palm	140
Fragrant Litsea	136
Fraxinus L.	240
Garcinia (see Lingnan Garcinia)	
Garcinia L.	
mangostana L.	82, 110
oblongifolia Champ. ex Benth.	**110–111**
Garrya Douglas ex Lindl.	56
Garryaceae	56
Glochidion (see Sri Lankan Glochidion)	
Glochidion J. R. Forst. & G. Forst.	112
hongkongense Müll. Arg.	112
littorale auct. non Blume	112
zeylanicum (Gaertn.) A. Juss.	**112–113**
Gmelina	114
Gmelina L.	114
chinensis Benth.	**114–115**
Goosefoot tree	206
Gordonia	10, 188
Gordonia Ellis	10, 188
anomala Spreng.	188
axillaris (Roxb. ex Ker Gawl.) D. Dietr.	188
Grewia L.	
microcos L.	160
Grossulariaceae	126
Guger tree (see Chinese guger tree)	
Gum (see Formosan gum, or Sweet gum)	
Gutta-percha	212
Guttiferae (see Clusiaceae)	
Hackberry (see Chinese hackberry)	
Hainan Annslea	42
Hamamelidaceae	16, 106, 132, 198
Hanging bell flower	102
Hawthorn (see Hong Kong hawthorn)	
Hemp	232
Heritiera (see Coastal Heritiera)	
Heritiera Aiton	116
littoralis Aiton	64, **116–117**
Hibiscus (see Sea Hibiscus)	
Hibiscus L.	
pernambucensis Arruda	118
tiliaceus L.	10, 19, **118–119**, 228
Hickory	100
Holly (see Chinese holly, or Panaceae holly)	
Homalium	120
Homalium Jacq.	120
cochinchinense (Lour.) Druce **120–121**, 210	
fagifolium Benth.	120
Hong Kong abacus plant	112
Hong Kong Camellia	70
Hong Kong hawthorn	196
Hong Kong Magnolia	152
Hong Kong orchid tree	58
Hong Kong Pavetta	174
Hops	232
Hornbeam	7

Horsetail 74

Horsetail tree 74

Huanghuali 86

Humulus L. 232

Hydrangeaceae 126

Ilex L. 122

    rotunda Thunb. **122–123**

Illiciaceae 124

Illicium L. 124

    angustisepalum A. C. Sm.

        16, 44, **124–125**, 152

    spathulatum auct. non Y. C. Wu 124

Incense tree 48

India-charcoal Trema 232

Iron olive 212

Itea 126

Itea L. 126

    chinensis Hook. & Arn. **126–127**

Iteaceae 126

Ivy tree 206

Jackfruit 54

Japanese date plum 94

Japanese lacquer tree 230

Juglandaceae 100

Juglans L. 100

Kaki 94

Kandelia 128

Kandelia Wight & Arn.

    candel auct. non (L.) Druce 128

    obovata Sheue, H. Y. Liu & J. W. H. Yong

        64, **128–129**

    rheedii auct. non Wight & Arn. 128

Kapok 60

Keteleeria Carr.

    fortunei (A. Murr.) Carr. 17

Kusamaki 186

Labiatae (see Lamiaceae)

Lacquer tree (see Japanese lacquer tree)

Lamiaceae 66, 114

Lance-leaved Sterculia 216

Lantau star-anise 124

Lauraceae 16, 80, 136, 138, 148, 150

Laurus L.

    camphora L. 80

Layia Hook. & Arn.

    emarginata Hook. & Arn. 164

Leucaena Benth.

    glauca (Willd.) Benth. 130

    loucocephala (Lam.) de Wit **130–131**

Lime 160

Linden 160

Lingnan Garcinia 110

Liquidambar (see Chinese Liquidambar)

Liquidambar L. 132

    formosana Hance 9, 10, 11, **132–133**, 279

Lirianthe Spach

    championii (Benth.) N. H. Xia & C. Y. Wu

        152

Litchi (see Lychee)

Litchi Sonn.

    chinensis Sonn. 16, 92

Lithocarpus Blume

    glaber (Thunb.) Nakai 84, **134–135**

    harlandii (Hance ex Walpers) Rehd. 134

Litsea (see Frangrant Litsea, Mountain-pepper

    or Pond spice)

Litsea Lam. 138

    citrata auct. non Blume 136

    cubeba (Lour.) Pers. 80, **136–137**

    glutinosa (Lour.) C. B. Rob. 136, **138–139**

    sebifera Pers. 138

Livistona R. Br.

    chinensis (Jacq.) R. Br. ex Mart. **140–142**

Longan 92

Lophostemon Schott

    confertus (R. Br.) Peter G. Wilson & J. T.

    Waterh. 15, **144–145**, 182

Loquat 104

Lychee 16, 92

Macaranga Thouars 146, 154

    tanarius (L.) Müll. Arg.

        10, 40, 98, **146–147**, 154, 202

Machilus (see Chekiang Machilus, Woolly

    Machilus, or Zhejiang Machilus)

Machilus Nees 16, 148, 150

    chekiangensis S. K. Lee **148–149**, 150

    longipedunculata S. K. Lee & F. N. Wei 148

    thunbergii Sieb. & Zucc. 148

    velutina Champ. ex Benth. 80, 136, **150–151**

Magnolia (see Hong Kong Magnolia)

Magnolia L.

    championii Benth. 44, **152–153**

pumila auct. non Andr. 152

Magnoliaceae 16, 152

Mallotus Lour.

    cochinchinensis Lour. 154

    paniculatus (Lam.) Müll. Arg.

        40, 46, 98, **154–155**, 202

Mallow (see Sea coast mallow)

Malvaceae 60, 116, 118, 160, 194, 216, 228

Mangifera L.

    indica L. 230

Mango 230

Mangosteen 82, 110

Mangrove (see Many-petaled mangrove)

Manilkara Adans.

    zapota (L.) P. Royen 212

Many-petaled mangrove 64

Maple (see South China maple)

Marlea Roxb.

    begoniaefolia Roxb. 38

Melaleuca L.

    cajuputi Roxb. 15

        subsp. *cumingiana* (Turcz.) Barlow

        156–159

    cumingiana Turcz. 156

    leucadendra auct. non (L.) L. 156

    quinquenervia auct. non (Cav.) S. T. Blake

        156

Melia-leaved Evodia 224

Melicope J. R. Forst. & G. Forst.

    pteleifolia (Champ. ex Benth.) T. G. Hartley

        224

Mespilus L.

    japonica Thunb. 104

Microcos 160

Microcos L. 160

    nervosa (Lour.) S. Y. Hu **160–161**

    paniculata auct. non L. 160

Millett's Adinandra 36

Moraceae 16, 54, 108

Morris's persimmon 94

Mountain fig 108

Mountain-pepper 136

Mountain tallow 202

Myrica L. 162

    rubra (Lour.) Sieb. & Zucc. 34, **162–163**

Myricaceae 34, 162

Myrobalan 180

Myrsinaceae 52

Myrtaceae 16, 144, 156, 222

Mytilaria Lecomte 106

Nephelium L.

    longana Cambess. 92

Nettle tree (see Chinese nettle tree)

New Year flower (see Chinese New Year flower)

Oak (see Champion's oak, or Chestnut oak)

Oleaceae 166, 240

Olive (see Iron olive)

Orchid tree (see Hong Kong orchid tree)

Ormosia (see Emarginate-leaved Ormosia, or

    Shrubby Ormosia)

Ormosia Jacks. 164

    emarginata (Hook. & Arn.) Benth.

        86, **164–165**

Osmanthus (see Taiwan Osmanthus)

Osmanthus Lour. 166

    fragrans (Thunb.) Lour. 166

    matsumuranus Hayata **166–167**

Palaquium Blanco

    gutta (Hook.) Baill. 212

Paliurus Mill. 168

    aubletia Schult. 168

    ramosissimus (Lour.) Poir. **168–169**

Palm (see Chinese fan palm, or Fountain palm)

Palmae (see Arecaceae)

Panaceae holly 122

Pandanaceae 17, 170

Pandanus 170

Pandanus Parkinson

    tectorius Parkinson 10, **170–173**

Paper-bark tree 156

Pavetta (see Hong Kong Pavetta)

Pavetta L.

    hongkongensis Bremek. **174–175**

    indica auct. non L. 174

Pear (see also Callery pear, or Wild pear) 192

Pecan 100

Pentaphylacaceae 36, 42, 176

Pentaphylax (see Common Pentaphylax)

Pentaphylax Gardn. & Champ. 176

    euryoides Gardn. & Champ. 42, **176–177**

Persea Mill. 148

    americana Mill. 148

longipedunculata (S. K. Lee & F. N. Wei)
   Kosterm. 148
   velutina (Champ. ex Benth.) Kosterm. 150
Persimmon (see also Morris's persimmon) 94
Phoberos Lour.
   chinensis Lour. 210
Photinia (see Bentham's Photinia)
Photinia Lindl. 178
   benthamiana Hance 104, **178–179**, 196
Phyllanthaceae 112, 180
Phyllanthus L. 180
   emblica L. **180–181**
Pilular Adina 34
Pinaceae 17, 182, 184
Pine (see Buddhist pine, Chinese red pine, or
   Slash pine)
Pinus L. 17, 182
   elliottii Engelm. 9, 17, **182–183**, 184
   massoniana Lamb.
      9, 13, 14, 15, 17, 19, 72, 144, 182, **184–185**
   sinensis Lamb. 13
Pithecellobium Mart.
   lucidum Benth. 50
Podocarpaceae 186
Podocarpus L'Hert. ex Pers.
   macrophyllus (Thunb.) Sweet 17, **186–187**
Poison ivy 230
Polyspora Sweet 10, 188
   axillaris (Roxb. ex Ker Gawl.) Sweet
      10, 19, **188–189**, 190
Pond spice 138
Popinac (see White popinac)
Poplars 210
Populus L. 210
Portia tree 228
Prickly ash 240
Pyrenaria Blume
   spectabilis (Champ.) C. Y. Wu & S. X. Yang
      190–191
Pyrus L. 192
   calleryana Decne. **192–193**, 196
   communis L. 192
Quercus L. 72, 84
   subgenus *Cyclobalanopsis* (Oerst.) Schneid.
      84
   championii Benth. 84

   fissa Champ. ex Benth. 72
   thalassica Hance 134
Red pine (see Chinese red pine)
Red sandalwood 32
Reevesia 194
Reevesia Lindl.
   thyrsoidea Lindl. **194–195**
Rhamnaceae 168
Rhaphiolepis Lindl.
   indica (L.) Lindl. ex Ker **196–197**
Rhizophora L.
   gymnorhiza L. 64
Rhizophoraceae 64, 128
Rhodoleia 198
Rhodoleia Champ. ex Hook.
   championii Hook. **198–199**
Rhus L. 230
   chinensis Mill. **200–201**, 270
   succedanea L. 230
Rosaceae 104, 178, 192, 196
Rose apple 16, 222
Rosewood (see Brazilian rosewood, and South
   China rosewood)
Rottlera Willd.
   paniculata (Lam.) A. Juss. 154
Roxburgh's Engelhardia 100
Rubiaceae 34, 174
Rutaceae 30, 224, 240
Salicaceae 120, 210
Salix L. 210
Sandalwood (see Red sandalwood)
Sapindaceae 28, 92
Sapium Jacq.
   discolor (Champ. ex Benth.) Müll. Arg.
      40, 98, **202–203**
Sapotaceae 16, 212
Saurauia 204
Saurauia Willd. 204
   tristyla DC. **204–205**
Saxifragaceae 126
Saxifragales 126
Scarlet Sterculia 216
Schefflera J. R. Forst. & G. Forst.
   heptaphylla (L.) Frodin 10, 16, **206–207**
   octophylla (Lour.) Harms 206
Schima 208

Schima Reinw. ex Blume

    noronhae auct. non Reinw. ex Blume   208

    superba Gardner & Champ.   16, **208–209**

Schisandraceae   124

Scolopia (see Chinese Scolopia)

Scolopia Schreb.

    chinensis (Lour.) Clos   120, **210–211**

    crenata auct. non Clos   210

    saeva (Hance) Hance   210

Screw pine   17, 170

Sea coast mallow   118

Sea Hibiscus   118

Sea mango   78

Sebifera Lour.

    glutinosa Lour.   138

She-oak   74

Shrubby Ormosia   164

Sichuan pepper   240

Sideroxylon L.   212

    wightianum Hook. & Arn.   212

Silver-back Artocarpus   54

Sinosideroxylon (see Wight's Sinosideroxylon)

Sinosideroxylon (Engl.) Aubrév.   212

    wightianum (Hook. & Arn.) Aubrév.

        212–213

Slash pine   182

Sloanea (see Chinese Sloanea)

Sloanea L.   96, 214

    hongkongensis Hemsl.   214

    sinensis (Hance) Hemsl.   **214–215**

Snow-bell (see Cork-leaved snow-bell)

South China maple   28

South China rosewood   86

Sparrmanniaceae   160

Sponia Comm. ex Decne.

    velutina Planch.   232

Sri Lankan Glochidion   112

Staphylea L.   234

Staphyleaceae   234

Star-anise (see Lantau star-anise)

Sterculia (see Lance-leaved Sterculia, or Scarlet

    Sterculia)

Sterculia L.

    lanceolata Cav.   16, 194, **216–217**

Stillingia L.

    discolor Champ. ex Benth.   202

Strawberry tree   162

Stylidium Lour.   19

    chinense Lour.   20

Styracaceae   218

Styrax L.   218

    confusus Hemsl.   218

    odoratissimus Champ. ex Benth.   218

    suberifolius Hook. & Arn.   **218–219**

Sugar-apple   44

Sumac   200

Sweet Artocarpus   54

Sweet gum   132

Sweet-leaf (see Dense-flowered sweet-leaf)

Sweetsop   44

Sweet Viburnum   238

Symingtonia Steenis

    tonkinensis (Lec.) Steenis   106

Symplocaceae   220

Symplocos (see Congested Symplocos)

Symplocos Jacq.   220

    congesta Benth.   **220–221**

Syzygium Gaertn.

    jambos (L.) Alston   16, 144, **222–223**

Taiwan Acacia   26

Taiwan Osmanthus   166

Tallow (see Mountain tallow)

Tanoak   134

Tarrietia Blume   116

Ternstroemiaceae   176

Tetradium Lour.   224

    glabrifolium (Champ. ex Benth.) T. G.

    Hartley   **224–227**

Tetranthera Jacq.

    citrifolia Juss.   138

    polyantha Wall.   136

Theaceae   16, 36, 68, 70, 176, 188, 190, 208

Thespesia Sol.   228

    populnea (L.) Sol. ex Corr.   **228–229**

Thorny wing nut   168

Thymelaeaceae   48

Tiliaceae   160

Toxicodendron Mill.   230

    radicans O. Kuntze   230

    succedaneum (L.) O. Kuntze   **230–231**

    vernicifluum (Stokes) F. A. Barkley   230

Trema (see India-charcoal Trema)

Trema Lour.	232
amboinensis (Willd.) Blume	232
orientalis auct. non (L.) Blume	232
tomentosa (Roxb.) Hara	**232–233**
Triadica Lour.	
cochinchinensis Lour.	202
Tristania R. Br.	
conferta R. Br.	144
Tung-oil tree	236
Turn-in-the-wind	154
Turpinia	234
Turpinia Vent.	234
cochinchinensis auct. non (Lour.) Merr.	234
montana (Blume) Kurz	**234–235**
nepalensis auct. non Wall. ex Wight & Arn.	234
pomifera auct. non DC.	234
Tutcheria (see Common Tutcheria)	
Tutcheria Dunn	
spectabilis (Champ.) Dunn	190
Ulmaceae	76, 232
Ulmus L.	232
Varied-leaf fig	108

Verbenaceae	114
Vernicia Lour.	236
fordii (Hemsl.) Airy Shaw	236
montana Lour.	40, 98, 202, **236–237**
Viburnum (see Sweet Viburnum)	
Viburnum L.	238
odoratissimum Ker Gawl.	**238–239**
Walnut	100
Wattle (see Ear-pod wattle)	
Wax tree	230
White popinac	130
Wight's Sinosideroxylon	212
Wild pear	192
Willow (see also Yellow basket willow)	210
Wing nut (see Thorny wing nut)	
Wood-oil tree	236
Woolly Machilus	150
Yellow basket willow	100
Yellow cow wood	82
Zanthoxylum L.	
avicennae (Lam.) DC.	**240–241**
lentiscifolium Champ.	240
Zhejiang Machilus	148

—— ❧ ——

Autumn leaves of *Liquidambar formosana*

Index of Chinese Plant Names

七姐果	216	石筆木	190
八角楓	38	吊鐘	102
三年桐	236	吊鐘王	198
土沉香	48	朴樹	76
大果馬蹄荷	106	米花樹	206
大頭茶	188	老鼠刺	126
大嶼八角	124	耳果相思	24
山松	184	耳葉相思	24
山油柑	30	血桐	146
山香圓	234	豆梨	192
山烏桕	202	車輪梅	196
山黃麻	232	刺柊	210
山蒼樹	136	枇杷	104
中華杜英	94	油桐	236
五列木	176	金鳳	90
天料木	120	亮葉猴耳環	50
孔雀豆	32	南嶺黃檀	88
木油桐	236	恒春黃槿	228
木荷	208	春花	196
木麻黃	74	柯	134
木棉	60	洋紫荊	58
木薑子	136	珊瑚樹	238
木蠟樹	230	秋茄樹	128
木欖	64	紅皮	218
水東哥	206	紅皮糙果茶	68
水筆仔	128	紅花荷	198
水團花	34	紅苞木	198
牙香樹	48	紅膠木	144
牛矢果	166	降真香	30
牛耳楓	88	革葉鐵欖	212
牛尾松	74	香港大沙葉	174
凹葉紅豆	164	香港木蘭	152
台灣相思	26	香港茶	70
布渣葉	160	香港猴歡喜	214
白千層	156	香港算盤子	112
白木香	48	栓葉安息香	218
白桂木	54	桂圓	92
白楸	154	桃葉珊瑚	56
石栗	40	浙江潤楠	148
石梓	114	海杧果	78
石斑木	196	海紅豆	32

破布葉	160	裸花紫珠	68
茜木	174	銀合歡	130
茶梨	42	銀柴	48
馬甲子	168	銀葉樹	116
馬尾松	184	閩粤石楠	178
假蘋婆	216	鳳凰木	90
密花山礬	220	劍葉械	28
梭羅樹	194	樟	80
荷樹	208	潺槁樹	138
野杜英	96	餘甘子	180
野漆樹	230	鴨腳木	208
麻子梨	192	龍眼	92
猴歡喜	214	嶺南山竹子	110
番荔枝	44	嶺南青岡	84
絨毛潤楠	150	濕地松	182
華石梓	114	濱海械	28
華南飛蛾樹	28	簕欓	240
黃牙果	110	簕欓花椒	240
黃牛木	84	繖楊	228
黃杞	100	鵝掌柴	208
黃桐	98	羅浮柿	94
黃瑞木	36	羅傘樹	52
黃槿	118	羅漢松	180
愛氏松	182	鰲蒴錐	72
楊梅	162	鐵冬青	122
楓香	132	鐵欖	212
楝葉吳茱萸	224	露兜樹	170
鼠刺	126	變葉榕	108
蒲桃	222	鹽膚木	200
蒲葵	140		

—— ❦ ——

Flowering branch of *Enkianthus quinqueflorus*

Subject Index

———— ❧ ————

Main entries for each topic are arranged according to individual species entries where applicable.

Acorns
 Cyclobalanopis championii — 84
 Lithocarpus glaber — 134
Agarwood
 Aquilaria sinensis — 48
Allelopathy
 Acacia confusa — 26
 Viburnum odoratissimum — 238
Aluminium accumulation
 Daphniphyllum calycinum — 88
Androdioecy
 Saurauia species — 204
Androecium — 17
Androgynophore
 Reevesia thyrsoidea — 194
 Sterculia lanceolata — 216
Angiosperms — 17
Ant-plant symbioses
 Acacia auriculiformis — 24
 Aleurites moluccana — 40
 Diospyros morrisiana — 94
 Endospermum chinense — 98
 Gmelina chinensis — 114
 Macaranga tanarius — 146
 Sapium discolor — 202
 Vernicia montana — 236
Antimicrobial compounds in fruits
 Daphniphyllum calycinum — 88
Arils
 Acacia auriculiformis — 24
 Sloanea sinensis — 214
Bacterial symbiosis (see Leaf nodules, and Root nodules)
Binomials — 19
Biofuel
 Vernicia fordii — 236
 Vernicia montana — 236
Biogeographical disjunctions

 Liquidambar formosana — 132
Brunfels, Otto — 21
Buttresses (of trunk)
 Heritiera littoralis — 116
Calyx — 17
Capsules
 Aporosa dioica — 46
 Bombax ceiba — 60
 Camellia crapnelliana — 68
 Camellia hongkongensis — 70
 Cratoxylum cochinchinense — 82
 Enkianthus quinqueflorus — 102
 Exbucklandia tonkinensis — 106
 Hibiscus tiliaceus — 118
 Lophostemon confertus — 144
 Macaranga tanarius — 146
 Polyspora axillaris — 188
 Reevesia thyrsoidea — 194
 Schima superba — 208
Carpels — 17
 unfused
  Annona squamosa — 44
  Illicium angustisepalum — 124
  Magnolia championii — 152
Co-evolution (of plants and pollinators) — 18
Collinson, Thomas B. — 14
Corner, E. J. H. — 214
Corolla — 17
Cosmetics
 Aleurites moluccana — 40
 Litsea glutinosa — 138
Country Parks in Hong Kong — 15
Culinary uses
 Adinandra millettii — 36
 Aleurites moluccana — 40
 Artocarpus altilis — 54
 Artocarpus heterophyllus — 54
 Cinnamomum camphora — 80

Osmanthus fragrans	166	Endemic (to Hong Kong)		
Cultivars	20	*Bauhinia purpurea × variegata*	58	
Bauhinia purpurea × variegata	58	*Illicium angustisepalum*	124	
Pyrus calleryana	192	Epimatia		
Cupules		*Podocarpus macrophyllus*	186	
Castanopsis fissa	72	Epithet, specific	19	
Cyclobalanopsis championii	84	Extra-floral nectaries (see Nectaries)		
Lithocarpus glaber	134	Family name	20	
Daley, P. A.	15	Fasciclodes		
Da Vinci, Leonardo	21	*Cratoxylum cochinchinense*	82	
Deforestation of Hong Kong	13–16	*Feng shui* (see *Fung shui*)		
Dioecy		Fig wasps		
Aporosa dioica	46	*Ficus variolosa*	108	
Daphniphyllum calycinum	88	Fire-breaks		
Endospermum chinense	98	*Acacia confusa*	26	
Ilex rotunda	122	*Viburnum odoratissimum*	238	
Litsea cubeba	136	Fire resistance (adaptations)		
Litsea glutinosa	138	*Melaleuca cajuputi*	156	
Macaranga tanarius	146	*Rhaphiolepis indica*	196	
Mallotus paniculatus	154	Follicles		
Pandanus tectorius	170	*Illicium angustisepalum*	124	
Vernicia montana	236	*Magnolia championii*	152	
Dioecy, cryptic		*Sterculia lanceolata*	216	
Saurauia species	204	*Tetradium glabrifolium*	224	
Symplocos congesta	220	Frugivory (see Fruit dispersal)		
Drupes		Fruit chemistry		
Alangium chinense	38	*Acronychia pedunculata*	30	
Aucuba chinensis	56	*Adinandra millettii*	36	
Callicarpa nudiflora	66	Fruit dispersal (see also Seed dispersal)		
Celtis sinensis	76	by bats		
Daphniphyllum calycinum	88	*Syzygium jambos*	222	
Litsea cubeba	136	by birds		
Microcos nervosa	160	*Acronychia pedunculata*	30	
Osmanthus matsumuranus	166	*Adinandra millettii*	36	
Pandanus tectorius	170	*Ardisia quinquegona*	52	
Phyllanthus emblica	180	*Callicarpa nudiflora*	66	
Symplocos congesta	220	*Celtis sinensis*	76	
Toxicodendron succedaneum	230	*Cinnamomum camphora*	80	
Viburnum odoratissimum	238	*Daphniphyllum calycinum*	88	
Dürer, Albrecht	21	*Ilex rotunda*	122	
Durian theory	214	*Litsea cubeba*	136	
Edible fruits		*Litsea glutinosa*	138	
Annona squamosa	44	*Livistona chinensis*	140	
Dimocarpus longan	92	*Machilus chekiangensis*	148	

Machilus velutina	150	
Microcos nervosa	160	
Osmanthus matsumuranus	166	
Rhaphiolepis indica	196	
Rhus chinensis	200	
Toxicodendron succedaneum	230	
by civets		
Artocarpus hypargyreus	54	
Diospyros morrisiana	94	
Microcos nervosa	160	
Myrica rubra	162	
by macaques		
Artocarpus hypargyreus	54	
Garcinia oblongifolia	110	
by muntjacs (barking deer)		
Phyllanthus emblica	180	
by water		
Cerbera manghas	78	
Heritiera littoralis	116	
Paliurus ramosissimus	168	
Pandanus tectorius	170	
Thespesia populnea	228	
by wind		
Acer sino-oblongum	28	
Casuarina equisetifolia	74	
Engelhardia roxburghiana	100	
Fruit types (see also Acorns, Capsules, Drupes, Follicles, Legume pods, Multiple fruits, Nuts, Pomes, Samaras, Schizocarpic fruits, Syconia, and Syncarpic fruits)	18–19	
Fuchs, Leonhart	21	
Fung shui woods	16	
Adenanthera microsperma	32	
Aquilaria sinensis	48	
Artocarpus hypargyreus	54	
Cinnamomum camphora	80	
Dimocarpus longan	92	
Schima superba	208	
Syzygium jambos	222	
Fung shui uses		
Podocarpus macrophyllus	186	
Galla chinensis		
Rhus chinensis	200	
Galls		

Aporosa dioica	46	
Rhus chinensis	200	
Genus name	19	
Gymnosperms	17	
Pinus elliottii	182	
Pinus massoniana	184	
Podocarpus macrophyllus	186	
Gynoecium	17	
Hinds, Richard Brinsley	13	
Hybrids	20	
Bauhinia purpurea × variegata	58	
Pyrus calleryana	192	
Hypanthium		
Eriobotrya japonica	104	
Lophostemon confertus	144	
Photinia benthamiana	178	
Rhaphiolepis indica	196	
Syzygium jambos	222	
Illustrations, botanical	20–21	
Inferior ovaries		
Aucuba chinensis	56	
Engelhardia roxburghiana	100	
Eriobotrya japonica	104	
Photinia benthamiana	178	
Insecticides		
Cinnamomum camphora	80	
Invasiveness		
Pyrus calleryana	192	
Lacquer		
Toxicodendron succedaneum	230	
Lateral thickening of stems (see Seconday growth)		
Latex		
Sinosideroxylon wightianum	212	
Leaf nodules (see Nodules)		
Legume pods		
Acacia auriculata	24	
Adenanthera microsperma	32	
Archidendron lucidum	50	
Leucaena leucocephala	130	
Ormosia emarginata	164	
'Little Hong Kong'	14, 198	
Lockhart, Stewart	14	
Mangrove trees		

Bruguiera gymnorhiza	64	Nitrogen-fixation (see Nodules)	
Heritiera littoralis	116	Nodules	
Kandelia obovata	128	on leaves	
Mattioli, Pier Andrea	21	*Ardisia quinquegona*	52
Medicines		*Pavetta hongkongensis*	174
Alangium chinense	38	on roots	
Aleurites moluccana	40	*Archidendron lucidum*	50
Callicarpa nudiflora	66	*Casuarina equisetifolia*	74
Cinnamomum camphora	80	*Myrica rubra*	162
Dimocarpus longan	92	Nomenclature	19–20
Melaleuca cajuputi	156	Nuts	
Microcos nervosa	160	*Castanopsis fissa*	72
Phyllanthus emblica	180	*Engelhardia roxburghiana*	100
Rhus chinensis	200	Oil, source of	
Zanthoxylum avicennae	240	*Aleurites moluccana*	40
Monocotyledons ('monocots')	17	*Melaleuca cajuputi*	156
Monoecy		*Vernicia fordii*	236
Aleurites moluccana	40	*Vernicia montana*	236
Celtis sinensis	76	Ornamentals	
Dimocarpus longan	92	*Bauhinia purpurea × variegata*	58
Glochidion zeylanicum	112	*Bombax ceiba*	60
Sapium discolor	202	*Celtis sinensis*	76
Sterculia lanceolata	216	*Delonix regia*	90
Trema tomentosa	232	*Enkianthus quinqueflorus*	102
Vernicia montana	236	*Podocarpus macrophyllus*	186
Multiple fruits	19	*Styrax suberifolius*	218
Mutualism (plant-pollinator)		Osbeck, Pehr (Peter)	13
Ficus variolosa	108	Ovules	17
Glochidion zeylanicum	112	Perianth, undifferentiated (see Tepals)	
Naming of plants (see Nomenclature)		Petals	17
Nectaries		Phyllodes	
extra-floral		*Acacia auriculiformis*	24
Acacia auriculiformis	24	*Acacia confusa*	26
Aleurites moluccana	40	Pine wilt disease	15–16
Diospyros morrisiana	94	*Pinus massoniana*	184
Endospermum chinense	98	Pinewood nematode (*Bursaphelenchus*	
Gmelina chinensis	114	*xylophilus*)	15
Macaranga tanarius	146	*Pinus massoniana*	184
Mallotus paniculatus	154	Pistillodes	
Sapium discolor	202	*Tetradium glabrifolium*	224
Vernicia fordii	236	*Zanthoxylum avicennae*	240
Vernicia montana	236	Plant fibres	
floral		*Bombax ceiba*	60
Machilus velutina	150	Pollen (adaptations for dispersal)	

Pinus elliottii	182	
Podocarpus macrophyllus	186	
Pollination		
by bats	18	
Syzygium jambos	222	
by bees	18	
Alangium chinense	38	
Castanopsis fissa	72	
Cinnamomum camphora	80	
Cratoxylum cochinchinense	82	
Diospyros morrisiana	94	
Elaeocarpus chinensis	96	
Enkianthus quinqueflorus	102	
Eriobotrya japonica	104	
Homalium cochinchinense	120	
Ilex rotunda	122	
Litsea cubeba	136	
Machilus chekiangensis	148	
Machilus velutina	150	
Pentaphylax euryoides	176	
Sapium discolor	202	
Schefflera heptaphylla	206	
Schima superba	208	
Styrax suberifolius	218	
Tetradium glabrifolium	224	
Toxicodendron succedaneum	230	
Trema tomentosa	232	
Viburnum odoratissimum	238	
by beetles	18	
Annona squamosa	44	
Viburnum odoratissimum	238	
by birds	18	
Bruguiera gymnorhiza	64	
Delonix regia	90	
Rhodoleia championii	198	
Syzygium jambos	222	
by butterflies	18	
Pavetta hongkongensis	174	
Reevesia thyrsoidea	194	
Schefflera heptaphylla	206	
Schima superba	208	
Viburnum odoratissimum	238	
by flies	18	
Lithocarpus glaber	134	
Schefflera heptaphylla	206	
Viburnum odoratissimum	238	
by hemipteran 'flower bugs'		
Macaranga tanarius	146	
by moths	18	
Glochidion zeylanicum	112	
by wasps		
Diospyros morrisiana	94	
Polyspora axillaris	188	
Schefflera heptaphylla	206	
Viburnum odoratissimum	238	
by wind	18	
Aporosa dioica	46	
Casuarina equisetifolia	74	
Cyclobalanopsis championii	84	
Engelhardia roxburghiana	100	
Liquidambar formosana	132	
Mallotus paniculatus	154	
Pandanus tectorius	170	
Pinus elliottii	182	
Pollination drops		
Pinus elliottii	182	
Podocarpus macrophyllus	186	
Pomes		
Eriobotrya japonica	104	
Photinia benthamiana	178	
Rhaphiolepis indica	196	
Prevention of self-pollination (see Dioecy, Monoecy, Protandry, and Protogyny)		
Principle of priority	20	
Prop roots		
Bruguiera gymnorhiza	64	
Protandry		
Delonix regia	90	
Protogyny		
Annona squamosa	44	
Machilus velutina	150	
Reafforestation of Hong Kong	15	
Reafforestation trees		
Acacia auriculiformis	24	
Acacia confusa	26	
Casuarina equisetifolia	74	
Leucaena leucocephala	130	
Lophostemon confertus	144	

Melaleuca cajuputi	156
Pinus elliottii	182
Pinus massoniana	184
Pyrenaria spectabilis	190
Schima superba	208
Trema tomentosa	232
Vernicia fordii	236
Vernicia montana	236
Ridley, H. N.	116
Roosting sites (for bats)	
Livistona chinensis	140
Root nodules (see Nodules)	
Samaras	
Casuarina equisetifolia	74
Scatter hoarding (of seeds)	
Camellia crapnelliana	68
Camellia hongkongensis	70
Castanopsis fissa	72
Cyclobalanopsis championii	84
Lithocarpus glaber	134
Pyrenaria spectabilis	190
Styrax confusus	218
Styrax odoratissimus	218
Schizocarpic fruits	
Acer sino-oblongum	28
Secondary growth (of stems)	16–17
Seed dispersal (see also Fruit dispersal)	
by birds	
Adenanthera microsperma	32
Aporosa dioica	46
Archidendron lucidum	50
Elaeocarpus chinensis	96
Macaranga tanarius	146
Ormosia emarginata	164
Podocarpus macrophyllus	186
Rhaphiolepis indica	196
Sloanea sinensis	214
Sterculia lanceolata	216
Tetradium glabrifolium	224
by civets	
Elaeocarpus chinensis	96
Myrica rubra	162
by macaques	
Artocarpus hypargyreus	54

Garcinia oblongifolia	110
by rodents	
Camellia crapnelliana	68
Camellia hongkongensis	70
Castanopsis fissa	72
Cyclobalanopsis championii	84
Lithocarpus glaber	134
Pyrenaria spectabilis	190
by wasps	
Aquilaria sinensis	48
by water	19
Hibiscus tiliaceus	118
Kandelia obovata	128
by wind	19
Adina pilulifera	34
Bombax ceiba	60
Cratoxylum cochinchinense	82
Exbucklandia tonkinensis	106
Lophostemon confertus	144
Melaleuca cajuputi	156
Pentaphylax euryoides	176
Polyspora axillaris	188
Reevesia thyrsoidea	194
Rhodoleia championii	198
Schima superba	208
Seed dormancy	
Viburnum odoratissimum	238
Seemann, Berthold	13, 70
Semi-inferior ovaries	
Anneslea fragrans	42
Sepals	17
Spadix	
Pandanus tectorius	170
Spathe	
Pandanus tectorius	170
Species name	19
Specific epithet (see Epithet)	
Sporangia	17
Sporophylls	17
Stamens	17
bisporangiate	
Exbucklandia tonkinensis	106
Staminodes	
Litsea cubeba	136

Litsea glutinosa	138
Machilus velutina	150
Tetradium glabrifolium	224
Zanthoxylum avicennae	240

Sugar content of fruits (see Fruit chemistry)

Superior ovaries

| *Pentaphylax euryoides* | 176 |

Syconia

| *Ficus variolosa* | 108 |

Syncarpic fruits

| *Artocarpus hypargyreus* | 54 |
| *Ficus variolosa* | 108 |

Synchronous flowering

| *Delonix regia* | 90 |
| *Dimocarpus longan* | 92 |

Synonyms	20
Tai Tam Tuk	14
Talbot, L. M. & M. H.	15

Tepals

Illicium angustisepalum	124
Machilus velutina	150
Magnolia championii	152

Timber

Aleurites moluccana	40
Cinnamomum camphora	80
Dalbergia nigra	86
Dalbergia odorifera	86
Dimocarpus longan	92
Exbucklandia tonkinensis	106
Lophostemon confertus	144
Pinus massoniana	184

Toxic plants

Aleurites moluccana	40
Cerbera manghas	78
Rhus chinensis	200
Toxicodendron succedaneum	230
Viburnum odoratissimum	238

Urushiol

| *Toxicodendron succedaneum* | 230 |

Uses (see Cosmetics, Culinary uses, Edible fruits, *Fung shui*, Fire-breaks, Insecticides, Lacquer, Medicines, Ornamentals, Plant fibres, and Timber)

Vivipary

| *Bruguiera gymnorhiza* | 64 |
| *Kandelia obovata* | 128 |

—— ❧ ——

Fruit pods of *Acacia confusa*

About the Authors

Sally Grace Bunker, a former school principal, is a Fellow of the Society of Botanical Artists (UK) and is currently focusing on recording the tree flora of Hong Kong in detailed botanical watercolour illustrations.

Richard M. K. Saunders is a Professor in the School of Biological Sciences at The University of Hong Kong. His research focuses on the diversity and evolution of flowering plants in Asia.

Pang Chun Chiu is currently a Post-doctoral Fellow at The University of Hong Kong. His research interests focus on plant ecology and ecological restoration.

www.ingramcontent.com/pod-product-compliance
Lightning Source LLC
Chambersburg PA
CBHW080901030426
42335CB00019B/2416